Technology and Employment

Technology and Employment

Concepts and Clarifications

Eli Ginzberg,
Thierry J. Noyelle,
and Thomas M. Stanback, Jr.

Westview Press / Boulder and London

223942

331.1
G 492

Conservation of Human Resources Studies in the New Economy

Published in 1986 in the United States of America by Westview Press, Inc.; Frederick A. Praeger, Publisher; 5500 Central Avenue, Boulder, Colorado 80301

Library of Congress Cataloging-in-Publication Data
Ginzberg, Eli, 1911–
  Technology and employment.
  (Conservation of Human Resources studies in the new economy)
  Includes index.
  1. Labor supply—United States—Effect of technological innovations on.   2. Women white collar workers—United States—Effect of technological innovations on.   3. Microelectronics—Social aspects—United States.   I. Noyelle, Thierry J. II. Stanback, Thomas M.   III. Title.   IV. Series.
  HD6331.2.U5G55   1986   331.1′0973   86-5556
  ISBN 0-8133-0399-0 (alk. paper)

Composition for this book was provided by Conservation of Human Resources.

Printed and bound in the United States of America

The paper used in this publication meets the requirements of the American National Standard for Permanence of Paper for Printed Library Materials Z39.48-1984.

6   5   4   3   2   1

Contents

Tables

Preface

This volume is the first of four publications that will present the research on technology and employment carried out by Conservation of Human Resources of Columbia University over the past several years. This research was started with a small grant from the Rockefeller Foundation in 1982. The original grant was renewed in 1983 by Dr. Bernard Anderson, then head of the Social Sciences Division of the Rockefeller Foundation, so that we could complete the case studies of selected large corporations, which we had started as a way to understand the impacts of the new computer technology on both organizational structures and employment patterns.

In 1983, Conservation of Human Resources, encouraged by the progress made possible by the Rockefeller grant, approached the Ford Foundation for more substantial funding. Our objective was to explore more extensively, through case studies of various sectors, the impact of the computer-communications revolution on the employment of different groups of workers, especially women and minorities, who might be particularly vulnerable to the changes under way. In the summer of 1985, we received additional modest funding from the Ford Foundation to enable us to complete the writing of the four emerging monographs. In our relations with the Ford Foundation, we have had the benefit of sympathetic understanding and critical assistance from Susan Berresford, vice president for United States and International Affairs Programs, and Gordon Berlin, program officer, which we gratefully acknowledge.

During the course of our research, we profited from periodic discussions with two of our colleagues, Thomas Bailey and Anna Dutka, which we gratefully acknowledge.

The core members of the Conservation staff who collaborated on these studies consisted of Dr. Thierry Noyelle and Professor Thomas M. Stanback, Jr., with my continuing involvement in selected phases of the planning, analysis, and writing. For this book, Dr. Noyelle wrote Chapters 4 and 6; Professor Stanback is the author of Chapters 2 and 5; and I wrote Chapters 3 and 7. The three of us collaborated on Chapter 1. I also assumed most of the editorial task of preparing the manuscript for publication. Despite the foregoing specifications, the book is an integrated product for which we assume joint responsibility. The ideas that we present and the conclusions that we reach are the outcome of a collegial process.

Chapter 1 is based on one of our progress reports to the Ford Foundation and represents an attempt to conceptualize what we have learned about the pace and impact of the new technology on employment, particularly on the employment of women and minorities. Although the chapter provides selective empirical insights from our field investigations, its thrust is on conceptualization. Our three other monographs will provide in-depth accounts of these field studies.

Chapter 2, "Work Force Trends," provides a bridge from the broader aspects of the subject presented in Chapter 1 to the more specific focus of Chapters 3 through 6. The detailed employment data presented in Chapter 2 provide the reader with a firm foundation from which to explore and assess changes that the new technology is bringing in its wake.

Chapters 3 and 4 were written at the specific request of the Panel on Technology of the Committee on Women's Employment of the National Academy of Sciences. The chapters deal with separate, but related, aspects of the impact of the new technology on female workers. Specifically, they analyze the extent to which established approaches to Equal Employment Opportunity (EEO) may need to be modified in light of changes brought about by the new technology. A major analytical linkage between the two chapters is the increasing determination of employers to hire new workers with more capacities and competencies and the vulnerability and danger that this poses to less-educated women and minorities.

Chapters 5 and 6 draw heavily on earlier work carried out by the Conservation of Human Resources that studied the hierarchy of cities in the United States and focus on the likely impact of the computer-communications revolution on patterns of employment in different locations. Chapter 5 looks at the system of cities, and Chapter 6 considers some of the threats that New York City may face in the years ahead as new linkages between the computer and communications permit employers to tap into new labor supplies, including foreign ones, at the clerical and technical level as well as among senior professionals and engineers. The chapter also points to the challenges that New York City banks may encounter in responding to the financing needs of small- and middle-sized enterprises, which in the past have not been on their priority list of customers. It concludes with observations as to the dangers that will result from the large number of minority youth who stop attending school before acquiring a high school diploma, which has become the minimum requirement for employment in most business service enterprises on which New York City's recovery has been based.

In Chapter 7, we consider directions for policy, including existing mechanisms and instrumentalities as well as new ones that should be put in place so that the considerable number of workers, old and new, whose employability is threatened by the new technology have a better chance of finding and keeping jobs in the years ahead. The alternative to strengthening the labor market infrastructure and linking it more effectively to the educational system is to turn our backs on improving job prospects for the disadvantaged and the displaced and to restrict ourselves to assisting them through welfare and other forms of income transfers. As our society continues to value the work ethic, keeping large numbers of displaced and disadvantaged workers on the dole for decades should not be our policy of choice.

Eli Ginzberg, Director
Conservation of Human Resources,
Columbia University

Technology and Employment

1

A Pass at Conceptualization

Introduction: A New Technological Era

Four principal questions are raised at the outset of this chapter: (1) What makes the current phase of computerization qualitatively different from that which preceded it? (2) How can we assess the pace and scope of diffusion of the new technology? (3) What are the determinants of the pace and scope of diffusion? (4) What is the impact of the new technology on employment, especially on the employment of women and minority workers who have traditionally been at a disadvantage?

In order to assess the nature of the transformations that are being brought about by the new generation of microprocessing and distributed data processing technology, it is helpful to contrast the emerging era with the one that preceded it and that began with the introduction of the first commercial computers in the late 1950s.

The first stage of commercial computer technology was primarily directed at putting in place large number-crunching machines that were able to do a great many calculations at high speed and at steadily lower unit cost. The mainframe computer became the recordkeeping war-horse for large companies, and the data processing service bureaus performed many of the same tasks for a great number of middle- and small-sized companies. This new number-crunching capability provided U.S. businesses with a much-improved way to stay on top of many administrative functions, especially those with a heavy clerical content, from order taking and recordkeeping to billing, accounts receivable, and general ledger.

In brokerage houses, for example, the new capability for data processing was the only way the vastly enlarged burdens of the "back offices" could have been successfully handled after the early 1960s, when the volume of transactions carried out on the various exchanges increased spectacularly. AT&T was able to cope with large-scale expansion in areas of toll calculations, billing, and recordkeeping only because of labor economies made possible through new technology. AT&T had pointed out earlier that had it not been for the constant advances in technology it would have had to hire all the clerks who could work in order to cope with its increased volume.

As computerization proceeded, other important developments beyond the simple substitution of computers for clerical labor in routine data processing began to unfold. Among the most important developments was the computerization of inventory control procedures, followed later by crude linkages between electronically recorded point-of-sales information and inventory control and reordering mechanisms. These developments were facilitated by the emergence of on-line technology, making possible continuous data entry and direct monitorization at the point of operation. By the end of this first stage of computer technology (late 1960s and early 1970s), significant realignments of functions were beginning to occur, especially regarding the ways in which management was able to control much larger volumes of information and to perform critical operations not only more cheaply but much more effectively.

By the mid–1970s, a new era of computerization had been launched, marked by the rapid and widespread development of on-line distributed data processing systems involving improved data communication, massive data storage, high-speed processing capability, and the development of user friendly software. With this new technology, end users learned that data processing no longer needed to be centralized and performed at points far removed from operations but that computerized operations could be performed throughout an organization. In the process, users also discovered that the new technology could be utilized to handle a vast array of tasks, including file management (personnel files, vendor files, insurance policy files, welfare check files, etc.), simple decision-making procedures (file updating, insurance policy rating, order taking, etc.), and analyses required to assess operating efficiencies and to help plan strategies. In short, the computer is now the critical instrument used to restructure work throughout an organization from managerial decisionmaking to routine operations. No longer is the computer restricted largely to the improved processing of masses of data with only peripheral effects on the core work of organizations.

The Pace and Scope of Technological Change

In the process of being introduced far more pervasively than it has been earlier, the new technology challenges existing institutional arrangements and opens up opportunities for critical transformations in the following areas: work, organizations, markets, products, and private/public interfaces.

In effect, technological diffusion can be viewed as the outcome of the dialectical relationship between the new technology and the critical institutional arrangements underlying these five areas of transformation. On the one hand, the technology alters traditional institutional arrangements. On the other, older institutional arrangements moderate the speed with which the new technology diffuses.

The Transformation of Work

One of the greatest challenges brought about by the introduction of the new technology in the workplace is the opportunity it opens to create new operating procedures and new approaches to the division of labor. As one manager in charge of installing new systems in several agencies for the City of New York put it: "We must not simply use the technology to replicate the old manual procedures but seize the challenge of completely reorganizing the work flow."

Two examples illustrate this point. The City of New York is currently developing a full geo-coded file of the city's commercial, industrial, and residential structures. With this system, each building will receive a file number that can then be related directly to decentralized data bases maintained in various departments including, for example, information on real estate tax payment or tax delinquency and information on housing or fire violations. In the short run, by making available on-line information based on existing records, agencies will be able to double-check and update files as soon as new information is entered, which will eliminate the enormous redundancies involved with current manual procedures. For example, five similar complaints phoned by five tenants in the same building cannot currently be cross-checked. One immediate result of the new system is that it will be possible to avoid sending several inspectors to the same building to check on the same violation. In the long run, the new system will enable inspectors to collate various types of information on a specific building rather than being limited in their response by fragmented information. In addition, the new system will transform much of the work carried on in the head offices of agencies because much of the file updating will be handled directly by inspectors in the field. The inspectors will be able to update files via portable computerized units that can be downloaded to central data files by pay phones or from field offices.

A second example is drawn from the Corporate Electronic Fund Transfer department of a large bank in the midst of a major overhaul. Under the old system, the unit was organized around a traditional step-by-step division of work involving a half dozen work stations and a great deal of paper shuffling from one work station to the next. A call

that came in from a customer firm's treasurer for a specific transfer
was recorded by the clerk who answered the phone, who then passed
on the information to several additional work stations for validation,
revalidation, entry into the computer, and execution of the transaction.
Under the new system, the bank set up an experimental work station
where the simple decisionmaking procedures had been computerized
and a single clerk had the responsibility of managing a number of
customer accounts, receiving the original phone call, double-checking
the order, entering the data, and executing the transfer. The bank also
extended limited on-line access to the customer, permitting the initiation
of direct remote transactions and holding the clerk at the bank responsible
for security checks, transaction activations, and day-to-day customer
relations with the clients' treasurer's offices.

We encountered a large number of such examples in many different
settings. Their implications are profound. In many cases, they bring to
an end the classic manner in which most large manufacturing and
service industries have traditionally organized their work flows, based
on an extreme division of labor, of which the assembly line had been
the prototype. The transition from the old division of labor to the new
must contend with a number of obstacles and much resistance, for, as
noted below, it has major unsettling effects on the ways both managers
and workers have hitherto conceptualized their work and their inter-
dependencies and have performed their distinctive tasks.

The Transformation of Organizations

As noted earlier, during the first phase of computerization, the
introduction of computer services in most large organizations proceeded
as an add-on activity. In largely independent divisions, specialists in
charge of the new machines interacted infrequently with the rest of the
organization, except for units directly engaged in the many clerical
operations to which they were providing assistance. The new era of
distributed data processing renders this sort of ad hoc, peripheral
organizational adaptation dysfunctional; it calls instead for major organizational restructuring and often requires a fundamental revamping
of the lines of authority and decisionmaking power. Here again a few
examples will help capture the magnitude of potential changes that
confront firms and organizations.

In banking, for example, sometime more than a decade ago, managers
of a few firms, such as Citicorp, perceived that the new technology had
the potential to enable them to fundamentally alter the way their
institutions transacted business with key customer groups, ranging from
the neighborhood depositor to the large corporate borrower, and in the

process to step out front and gain a significant lead over competitors. Data made available by Citicorp suggest that in consumer branch banking alone, the bank invested over $1 billion in the mid- and late 1970s in building a new technological infrastructure before the system yielded any profits.

The important point is that in the process of making broader and fuller use of the new technology, Citicorp managers rapidly came to understand that they had to alter the corporation's basic organizational structure. They did so by setting up three major independent units—consumer banking, corporate banking, and investment banking—each as its own profit center and each competing with the others for capital resources. In addition, Citicorp found it necessary to alter many of its traditional human resources policies: raising standards for new hires, putting a greater emphasis on entrepreneurially oriented middle managers, and altering its reward structures to pay for performance. To the outsider, the many changes often appeared to involve a mad scramble among managers, with different managers competing at times with similar products for the same consumer and individual units building their own computerized infrastructure with little attention to what other divisions were doing. Nevertheless, this "force-fed" decentralization of power helped Citicorp to embrace the new technology, to reposition itself in domestic and international markets, to develop a host of new or improved products, and ultimately to take the lead in redefining and altering the regulatory climate under which banking had traditionally operated.

The experience of the Bank of America stands in sharp contrast to that of Citicorp. At the time that Citicorp was planning its move into the new technological era, Bank of America led the U.S. banking industry in terms of total assets. But it substantially failed to embrace the new technology and make the necessary organizational and decisionmaking changes, which led to serious and continuing slippage in its profits and market position as Citicorp raced ahead. Furthermore, in its belated efforts to modernize, it now encounters much tougher competition for the simple reason that not only Citicorp but other more responsive banking organizations have taken the lead in many of Bank of America's traditional markets.

A similar story of changing organizational structures can be sketched in the not-for-profit sector. A key illustration is provided by New York City's government. The 1975 financial crisis demonstrated clearly that the City had become unmanageable if only because no one could tell for sure how much money was being spent on what, where, and when. At a minimum, a computerized system was needed to track the budgets of the City's 120 departments and agencies and to assure accounting rigor and financial controls. In order to accomplish these minimum goals, the City established a new agency—Financial Information Services Agency (FISA)—to take over the responsibility of clearing all purchases and payroll transactions and to execute payment. In addition, the City found it necessary to redefine budgetary responsibility and reallocate particular functions between operating and oversight agencies. In short, computerization and fiscal reform went hand in hand and brought additional pressures for further reorganization.

The Transformation of Markets

The new technology not only radically affects the cost and speed of information transfers, but it also has a major impact on the structure and functioning of markets. One result is that the new technology makes possible the manufacturing, marketing, and distribution of far more customized outputs than were heretofore possible. Rising consumer incomes have led to an increasing interest in quality, style, and variety, and manufacturers must respond to these demands in order to stay competitive. On the production side, the new technology permits the cost-effective operation of much smaller units than formerly was possible; witness the mini-steel mills. In addition, the computer enables the producer to customize output to the needs of a more segmented market and in turn to spend more effort on identifying and exploring new market niches. The combined effect of these trends is the altered behavior of both buyers and sellers in national markets and the restructuring of many firms, especially as regards employment and the size and nature of assets.

But perhaps the greatest potential of the new technology is in accelerating the integration of markets on a world scale. Within the past few years the new computer-communications technology has facilitated the integration of financial markets to the extent that for the first time in history we are seeing the emergence of a worldwide capital market. Trading in the various stock, commodities, and currency markets now goes on around the clock. On Wall Street, for example, it is possible for small firms that specialize in currency and interest swaps to carry out a profitable business by harvesting 1/32 of a percentage point of the value of the swaps!

Consider also the fact that until recently the major linkages between the less developed countries (LDCs) and the developed countries were largely limited to the trading of raw materials and manufactured goods, some branch-based production in LDCs by producers from developed countries, and some export of capital to LDCs. The new pattern of internationalization is increasingly one in which the large corporation seeks to operate more aggressively on a worldwide basis by optimizing the total use of its resources, including its use of professional resources, in terms of what it buys, makes, and sells and where it undertakes these several activities. The most striking development in this respect is the generalization of a model once pioneered by IBM, in which large corporations now attempt to make use of specialized talent pools in several countries simultaneously and to reorganize their engineering, R&D, and other professional functions in order to integrate these resources more effectively. For example, one large U.S. computer company has just started an advanced engineering design unit in Israel that will operate in close linkage with the firm's other R&D facilities located respectively in Western Europe and the United States. Likewise, Bechtel is now operating an engineering-drafting facility in India, and one of the largest U.S. financial firms farms out some of its software and systems development to an Indian consulting service firm.

The Transformation of Products

One test of the potency of a new technology is to assess its contribution to altering the nature of outputs. On this basis, the new technology has already established its claim, though more changes certainly lie ahead.

We noted earlier the large numbers of new products that the more advanced financial service firms have been able to offer different customers, from the major corporations (e.g., cash management accounts) to the small depositor (insured money rate accounts), as well as improvements in existing products, such as providing most consumers 24-hour access to their accounts through Automated Teller Machines (ATMs).

Other examples are readily at hand. Consider the explosive growth of credit cards, which has already had such a profound effect on consumer behavior. These changes would have been inconceivable had banking continued to rely on a paper-shuffling technology. Along similar lines, air travel reservations, auto rentals, hotel reservations, and ancillary services that represent major support infrastructures in the conduct of the travel business have been radically transformed and improved by the new computer-communications technology.

That the computer is leaving its mark on every phase of the production-consumption cycle, from the research laboratory to the ways in which we buy and pay for finished goods and services, underscores the fact that the new technology has already penetrated a great many sectors of the economy. It will, in time, penetrate further, leading to many new functions, new products, and new services that cannot even be imagined much less delineated as of today. One need point only to the burgeoning transformations occurring in two classic fields of human endeavor—

medicine and education—which are clearly just beginning to be affected by the new technology but where some of the changes are likely to be quite dramatic within the coming decade.

The Transformation of Private-Public Interfaces

This last area of change is particularly important to review in light of the present strong pro-market, pro-private sector bias that pervades the U.S. scene. Thus far the computer-communications industry has been heavily dependent on government initiatives, largely through defense expenditures. Most of the research and development funding, not only for computer hardware but also for training computer professional and technical staffs, can be traced back to the federal defense, nuclear energy, and space agencies. One conclusion is beyond doubt: In the absence of a large government market for the early mainframe computers the industry would have grown at a much slower pace.

Important as the federal government has been as financier of computer technology, it has not interfered directly with its development other than by interdicting sales of technologically sensitive equipment to unfriendly countries. On the other hand, the government did interfere in telecommunications by pushing for AT&T's divestiture in the belief that a telephone monopoly had lost its usefulness in the face of explosive developments in the communications industry and, in particular, increasing convergence between computer and communications technologies.

On a different front, the Department of Justice has recently ruled that cooperation by major computer companies in R&D activities was within anti-trust statutes so long as they did not collude at later stages of production and distribution. Within the past year or so, two such consortia organized by semi-conductor manufacturers have begun operations.

Clearly, important public policy issues remain unresolved. For example, the terms under which large firms, small firms, and individuals will be able to access the transmission channels that are being put in place as well as the associated questions of costs and confidentiality will almost certainly require some degree of regulation, both domestically and internationally. Moreover, the role of government in reaching agreements with other nations regarding such issues as transborder data flows is likely to increase dramatically in years ahead. In short, the notion that government is stepping or should step aside in this frontier area is misleading, although clearly the regulatory environment—both domestic and international—is being profoundly reshaped.

The Determinants of Pace and Scope of Diffusion

Six factors appear to be playing a major role in determining the pace and scope of technological change: competition, investment, organizational change, human resources, firm size, and technological infrastructure.

Competition

Enhanced competition has been a major force behind the adoption of new technologies and the development of the new work procedures that go along with their adoption. In the private sector, increasing internationalization since 1973 has put many firms and industries through the wrenching test of cost competition, especially labor cost competition with firms and industries in other nations where labor costs are often much lower. U.S. firms have often sought to respond to this challenge through purchasing new and better equipment, which allows them to reduce or eliminate their cost disadvantage, or through repositioning themselves vis-à-vis low-cost competitors in the marketplace. Opportunities for repositioning via the use of new technology have grown both because of the trend toward increasing market segmentation noted earlier and because the new technology itself has made it increasingly easier to customize outputs while keeping costs under control.

Deregulation, by spurring competition, has also played a significant role in speeding up the adoption of new technology by making it increasingly attractive for firms to begin to cross over into each others' markets (e.g., IBM and AT&T; Sears and Citicorp; etc.). In key industries, such as finance and telecommunications, both domestic deregulation and internationalization have propelled firms toward an increasing use of the new technology.

In the not-for-profit sector, it is primarily public pressure for cost containment that has spurred organizations in their search for ways to economize on labor costs through increasing recourse to the new technology. It should be noted, however, that these new cost-reducing pressures in the not-for-profit sector are recent (early 1980s), reflecting the increasing resistance of taxpayers and following upon the powerful competitive shocks that had an impact on the private sector in the mid-1970s. This timing helps to explain why the not-for-profit sector is only now beginning to use the new technology and doing so only selectively.

Investment

Despite the oft-heralded rapid decline in the cost of the new computer hardware, the magnitude of the investments typically called for in order to restructure a firm's operations and output may lead many executives to hesitate before gambling the future of their company on what is often an unchartered course.

Over the past six or seven years, the U.S. automobile industry has invested more than $50 billion in the development of new equipment, new processes, and new products in an effort to make up for years of inertia in responding to foreign competition. Yet $50 billion later, U.S. manufacturers have yet to demonstrate that they can succeed in combining the new technology and new products that will allow them to regain a solid footing in the market for small cars—now largely dominated by Asian producers. GM's Saturn project and Ford's and Chrysler's sister projects—all of which are based on fully automated production of "modular" cars—are directed to reestablishing a strong presence in this market, but the answer to whether such projects will succeed or fail is still several years and several billion dollars in the future.

We noted earlier the example of Citicorp's large investment to take the lead in the financial sector. Citicorp, however, is not alone. Such large capital outlays are increasingly common in the financial services sector where firms have not previously resorted to large-scale investment programs. Major financial organizations such as Merrill Lynch, American Express, Prudential, and CIGNA were reported in the early 1980s to be investing annually at rates of $200 million to $250 million each. Bank of America, a latecomer to the use of new technology, has announced that it is planning to invest $4 billion within the next five years.

The explanation for these large investments in the face of rapid decline in the cost of hardware lies in part in the proliferation of work stations, PCs, mini and mainframe computers, and, most important, in the explosive need for software, which often accounts for two-thirds or more of the new investments.

Problems of raising adequate investment funds are particularly difficult in the not-for-profit sector of the economy because access to new capital is limited. Accounting practices do not provide for depreciation on capital assets, and cash flow from this source cannot be counted on to help finance future investments. Most of the required funds must be raised from donors or from tax revenues.

Organizational Change

The process of technological change is also affected by the capacity of organizations to transform themselves so as to make full use of the opportunities that open up. Three key factors appear to act as constraints: generational differences, internal politics, and the organization's culture.

First, the ways that people react to a radically new technology depends in considerable measure on whether they have encountered it during their formative years. The older generation of managers is likely to view technology with apprehension and hostility. In contrast, the generation of young people entering the work force is likely to have much greater familiarity with computers, which will greatly facilitate its use and further diffusion.

Second, current technological change is based on the distribution of data and information processing capacity directly to users. This distribution requires a parallel decentralization of decisionmaking, which directly challenges the power of the highly centralized systems divisions established earlier to oversee the introduction of computers and computer applications. Furthermore, efforts to link large systems create new needs for data base management and for communication traffic management.

Finally, the new work procedures made possible by the new technology challenge the work and power of persons previously in critical positions. The capacity of the organization to alleviate the fears and respond to the concerns of those most directly threatened by change, while putting together a new culture that projects a positive image of the new technology, is a critical factor in determining the pace of future transformations. The capacity for "foot-dragging" on the part of individuals affected by change cannot be minimized.

Human Resources

Closely related to the problem of organizational change is the observation that the retraining of personnel, the reassignment of employees, the hiring of a new generation of workers, and the retirement of older personnel, are efforts that need time to be carried through. Furthermore, if one of the major lessons of the current period of transformation is that change rather than stability will increasingly become the norm, then firms and other user organizations face the major challenge of putting in place the institutional arrangements that will facilitate the continuing transition of workers from old jobs to new jobs. One large New York bank's new motto is that it can only afford to hire workers who are retrainable.

If a major factor determining the rate of diffusion is the capacity of an organization to alter the division of labor in ways appropriate to the new technology, it becomes critical that firms, industries, and the economy at large develop the capacity to rethink and reform the culture of work, the culture of organizations, and the culture of industrial and labor relations. Without such rethinking and actions, the pace of transition will be significantly slowed.

The presence of strong unions—particularly in the not-for-profit sector, where employment considerations loom particularly large—is likely to result in policies that require the employer to introduce the new technology without any loss of jobs (or income). Even in large nonunionized organizations, because of the rate of penetration and diffusion of the new technology and the potentially disruptive impact of large layoffs, few, if any, employees are let go. In short, successful change often requires that many workers be retrained and reassigned and only a few, at most, be dismissed.

Firm Size Factor

Several of our investigations—especially those in the financial sector, where we were able to study in detail not only a sample of large corporations but also a selected sample of medium- and small-sized institutions—clearly suggest that firm size is an important factor in the rate of diffusion.

Large organizations typically have the financial resources necessary not only to shoulder the cost of development and investment but also to pay for inevitable mistakes. This generalization does not ignore the fact that among large firms within the same industry some are changing faster than others or that some medium-sized or small firms may at times take a lead over larger competitors.

There have been numerous claims that the PC will significantly weaken the importance of organizational size by making extensive processing power accessible to smaller firms at low cost. Experience thus far does not seem to bear out this prediction. For the most part, smaller firms have used PCs (or minicomputers, whose cost has also declined dramatically) to bring back in-house the processing of accounting and inventory functions that previously had been contracted out to data processing firms or to larger competitors in their industry. Very few small firms have as yet ventured to use the PC to reorganize either their operational or managerial procedures, although such changes lie at the core of the successful application of new technology in larger organizations. In the aggregate, smaller firms tend to be "one step behind," and one must be careful not to underestimate the time that it may take to develop software that will enable these firms to restructure their operations.

The Technological Infrastructure

The concept of an infrastructure implies a system that has been designed in such a way that benefits can be shared by many users rather than restricted to a few. In the broadest sense one may think of the technological infrastructure in terms of human resources. The diffusion of training and know-how in the use and development of the new technology throughout society must be regarded as critical: It contributes to building a pool of skilled personnel from which ever larger numbers of firms and organizations can draw in developing new applications.

More narrowly defined, the technological infrastructure may be conceived of in terms of the apparatus that links together computerized subsystems. Advances in telecommunications will clearly play a role in accelerating the integration of currently freestanding systems. But the problems involved in building up this new infrastructure are far from trivial. Currently installed hardware often stands in the way as does the lack of suitable software. IBM's inability to bring to the market an operating local area network system, despite repeated claims that it is about to do so, points to some of the major problems that remain to be solved. Indeed, the European Economic Community's (EEC) Esprit program, in which the Community itself will expend billions of dollars of R&D monies to develop networking technology as a cooperative venture with a dozen computer developers, speaks to the formidable nature of the task that lies ahead.

A paradoxical development must be noted. Although the introduction of the microcomputer is contributing to enhanced diffusion, that very diffusion may act as a deterrent to the building of integrated systems. Confronted with a proliferation of subsystems utilizing diverse hardware, data bases, and software, firms have their work cut out for them as they seek new ways to speed integration.

The Time Dimension of Diffusion

In regard to the time dimension of diffusion, there appears to be a widespread confusion about the difference between the recognition that a new technology is ready to take off and the achievement of widespread diffusion throughout the economy. The latter, as we have seen, requires that a great many new institutional arrangements be put in place before the new technology can penetrate far and wide.

One way to assess the rate at which the new technology is likely to be diffused throughout the economy is to look back at the experience of the earlier era of computerization. Although there is no clear-cut division between the old era and the new, it can be postulated that old-style computing reached maturity by the mid–1970s or at the latest by the close of the 1970s. The beginning of that era can be dated back to the mid–1950s, when the IBM 650, the first commercially successful computer, was first introduced (1955). Rapid adoption followed immediately and accelerated throughout the 1960s, especially after the introduction of the IBM 1401, which, in retrospect, marked the turning point in the initial phase of computerization. Nevertheless, this first phase of computerization was still very much in progress well into the late 1960s and early 1970s. In fact, the interviews undertaken for our sectoral studies indicate that manual methods of processing large record systems are still in use in technologically backward organizations.

In looking back at that earlier era, one discovers that it required more than two decades for the extensive diffusion of electronic computerization to take place, although some of the employment impacts, as we shall see in the next chapter, were experienced much earlier.

The past five years (1980–1985) represent the turning point in the current phase of computerization. Our society has finally begun to understand and accept the idea that the new technology will radically alter the economy. Yet it must be remembered that the first experimentation with on-line distributed systems dates back to the late 1960s and early 1970s, when the insurance industry began to move aggressively to introduce large-scale on-line systems in back-office operations. That was fifteen years ago.

Grounds can be found for predicting that the future diffusion of the new technology will either be slower or more rapid than in the past. The hypothesis that it will speed up can be supported by the observation that the new and certainly more sophisticated generation of workers and managers are more open to the new technology. On the other hand, major obstacles in terms of costs, organizational change, and technical infrastructures support the view that diffusion will be protracted.

On balance, we find no strong evidence that the new era is unfolding either more or less rapidly than the former one. In any event, the United States is likely to be entering a period of sustained change during which the effects of computerization on work will be continuously broadened and deepened. These impacts are discussed below.

The Impact of Computerization on Employment

Several sets of major forces have brought about changes in employment and career opportunities during the postwar period, among which enhanced computer and telecommunications technology is only one. As a result, it is at times difficult to disentangle the impact of technology from that of other forces. Our field investigations focused on large organizations in selected sectors of the economy as a way to assess the impact of increased adoption of the technology on employment opportunities.

To gain some perspective regarding what has already taken place, it is useful to examine changes in the distribution of jobs among broad

TABLE 1.1
Occupation of Employed Workers (%), 1960–1982

	1960	1970	1975	1982
White-collar workers	43.3	48.3	49.8	53.7
Professional & technical	11.4	14.2	15.1	17.0
(teachers excl. college)	(2.5)	(3.4)	(3.6)	(3.3)
Managers & administrators	10.7	10.5	10.5	11.5
Sales workers	6.4	6.2	6.4	6.6
(retail trade)	(3.8)	(3.8)	(3.6)	(3.3)
Clerical workers	14.8	17.4	17.8	18.5
Blue-collar workers	36.6	35.3	33.0	29.7
Craft & kindred	13.0	12.9	12.9	12.3
Operatives	18.2	17.7	15.2	12.9
Non-farm laborers	5.4	4.7	4.9	4.5
Service workers	12.2	12.4	13.7	13.8
Farm workers	7.9	4.0	3.4	2.7
Services as % of U.S.				
nonagricultural employment	62.3	63.1	70.9	73.3

Source: U.S. Department of Commerce, Bureau of the Census, *Statistical Abstract of the United States 1984,* Washington, D.C., 1983.

occupational groups over three periods: the 1960s, the early and mid-1970s, and the late 1970s and early 1980s (1960–1970, 1970–1975, 1975–1982) (Table 1.1).

During the first half of the 1970s, although there was a continuing shift toward the services with their differentially higher clerical employment, the relative importance of clerical workers leveled off after a sharp rise during the 1960s. As noted previously, in the early 1970s the application of first- and second-generation computer technology to basic "number-crunching" tasks (accounting, inventory, billing, etc.) was reducing clerical labor requirements across a wide range of user organizations, even though the total volume of administrative paperwork was increasing at a very rapid rate. Computerization had spread rapidly in the 1960s, but its effects were not reflected in the aggregate data until the 1970s.

The latter 1970s and early 1980s point to yet another pattern of change. Clerical workers as a share of total employment increased slightly as did service workers and nonretail sales workers, while professional-technical and managerial workers experienced much larger relative gains. Of course, the data are too aggregative to permit detailed analyses of the new forces at work, but they suggest that significant occupational restructuring of the work force both reduced and expanded career opportunities for different groups of workers. Key findings regarding changes in employment patterns are presented below under three headings: the transformation of skills, the transformation of job access and mobility structures, and the displacement effects of the new technology.

The Transformation of Skills

An important initial observation is that the new phase of computerization does not lead ineluctably to downskilling, as has so often been argued. Instead, for many occupations the new technology often brings some degree of upskilling and an increase in the worker's range of responsibilities. Upskilling comes about for two principal reasons: first, because the most efficient use of the new technology leads to a reintegration of tasks that were once parcelled among many workers; second, because as intelligent systems take over processing functions, workers focus on diagnosis and problem-solving functions. Moreover, increased responsibility comes about because the scope of a worker's decisionmaking capacity (i.e., the range of the situations he or she is called upon to cope with) is typically expanded. For example, in the City of New York departmental payroll clerks working on the new system were called upon to execute a wider range of entries and checks to determine eligibility of payments than previously, not unlike the situation of the Corporate Electronic Fund Transfer department described earlier. In terms of increased responsibility, a typical example is that of claim settlement in insurance companies where improved computerized systems have placed more information at the clerks' disposal enabling them to settle larger amounts than in the past.

At the same time, however, the reorganization of the work process often renders redundant upper-level clerical and lower-level managerial personnel who had previously been responsible for overseeing a more subdivided work flow, bringing together information, and preparing critical analyses and reports. For example, in the insurance industry the work of policy raters and assistant underwriters is being rapidly taken over by the computer. Within universities, hospitals, and municipal governments, personnel previously assigned to the preparation of a variety of reports for top management, regulatory authorities, or others are becoming redundant as computerized data base management procedures are being put in place. The same generalization holds for many production activities, where first-line supervisors, traditionally responsible for quality control, find their skills made obsolescent by automated processing.

Overall, we found widespread evidence that as information becomes more readily accessed many upper-level clerical and technical workers and most professionals and managers are able to increase the scope and sophistication of their work. Many technical and professional workers find that the new computerization, by eliminating drudging "paperwork" and other chores, is freeing up their time and allowing them to focus on their major assignments and to perform at a higher level of expertise. This transformation is very much in evidence among buyers for large retail organizations, who are freed of enormously burdensome record-keeping activities by advanced computer systems and are thus able to focus more attention on merchandising tasks.

Finally, there is abundant evidence that much routine clerical work is being eliminated as the new systems bring about sharp reductions of paper handling. In insurance, for example, filing and messenger functions that were performed by large numbers of unskilled workers (typically youngsters out of high school) have been largely eliminated through computerization.

In light of these strong tendencies to reduce certain types of clerical work, it is helpful to review the data shown in Table 1.1 on clerical employment, which indicate a rise in the percentage of clerical employment. Explanation can be found in part in the sharp decline in blue-collar employment in the goods industries, which has led to a relative increase in the percentage of white-collar employment. In addition, the fact that the most menial clerical work is being eliminated by modern technology should not detract attention from the increasing emphasis on information processing throughout the economy. Clearly, clerical work, broadly defined, continues to be a major provider of jobs at the same time that there is upgrading within the clerical classification as a whole.

The above discussion has not treated a wide variety of low-level jobs that fall into such categories as service worker, sales worker, and laborer. Thus far, we have found little evidence that these workers have as yet been directly affected by computerization. One significant impact appears to come about through the capability of the computer to collect and analyze work flow information, thereby enabling the employers to expand their use of part-time employees. For example, department stores with modern point-of-sale computerization are able to estimate with considerable accuracy daily and weekly variations in requirements for sales clerks and thus can make greater use of part-time personnel. The same holds for balancing nursing personnel with hospital requirements. Moreover, the recordkeeping and scheduling involved for large numbers of part-time personnel are greatly facilitated by computerization.

*Transformation of Job Access
and Mobility Structures*

Changes in the demand for workers of differing skills are significantly altering opportunities for many individuals and groups to gain access to jobs and to move upward into desirable career paths. A major finding of our investigation is that the old hierarchical patterns of occupational advancement within large organizations (internal labor markets) is increasingly breaking down.

Historically, many workers tended to enter organizations at the "bottom of the ladder" and advance as they gained experience and acquired various company-specific skills. The new technology, however, tends to bring about a greater homogenization, or universalization, of skills (skills are less and less firm- or even industry-specific) and places more importance on external training, especially for many middle-level workers. The result is a growing delinking of jobs within the firm and industry. Less and less do workers within an organization move upward along traditional career paths; more and more, they enter the organization at various levels, depending on experience or formal training gained on the outside.

The significance of this development cannot be fully appreciated unless one takes account of changes in the educational patterns of the new generation of workers. The statistics are impressive: In 1960 slightly more than 10 percent of men and women between ages 25 and 29 had finished four or more years of college education; by 1980, the proportion had risen to nearly a fourth of this age group! This increase in the number of young men and women attending college and technical schools has put pressure on firms and nonprofit organizations to adjust their hiring procedures. In order to take advantage of this enlarged supply of young workers with superior educational backgrounds, employers are increasingly differentiating the job entry and career opportunities available.

These new arrangements have reduced the importance of traditional internal career ladders. One no longer becomes a buyer for a major retailing organization by starting out as a sales clerk; an insurance executive by starting out as a messenger; a bank branch manager by starting out as a teller. Almost all employees who rise to managerial and professional posts start as executive trainees after completing college or professional school.

Another factor contributing to the delinking of jobs for career advancement is the movement of back-office work away from headquarters to suburbs and to smaller cities and towns where labor (predominantly white women) is less costly and considered to be of a higher "quality" (in terms of literacy, less threat of union activity, and so on). Such geographical shifts of work make it even more unlikely for lower-level personnel to advance through the ranks. Moreover, widespread relocation of back-office work is having a significant negative impact on the employment of minorities in the inner-city.

This trend toward relocation of back-office work, facilitated by modern computer telecommunications systems, has been particularly evident among large banks, insurance companies, and credit card processing centers of the major oil companies, all of which were formerly located in the largest urban centers.

The upshot of these developments is that old career ladders are being eliminated and new patterns of job entry and career advancement are replacing them. There is still much to learn about these new labor market arrangements, but it appears that a worker's formal education and training, his or her ability to move among firms and industries or to gain additional knowledge and credentials outside the work place are becoming increasingly critical.

The Displacement Effects of the New Technology

Although our discussion thus far has been directed largely toward the qualitative impact of technological change, our detailed industry studies shed light on the issue of displaced workers and workers at risk in both services and goods industries and in the not-for-profit and profitmaking sectors.

To clear the deck of a simplistic but entrenched misconception, it seems inconsistent to argue on the one hand that the new technology is bringing about a jobless economy while witnessing on the other hand the net addition of 27 million new jobs between 1970 and 1985. But this observation is simply an aggregate assessment and may be misleading because it says little as to what is happening on an industry-by-industry basis. Once we turn to individual sectors, the picture that emerges is mixed but can be summarized in terms of two broad groups of industries. In the first group of industries, characterized by slowly growing or even shrinking demand, consisting mostly but not exclusively of older manufacturing industries, there has been large displacement of workers in recent years. Automobile and steel are prime examples, but to a lesser extent this displacement has also occurred in textile, insurance, and municipal government, which we studied in great detail.

Displacement in the auto and steel industries, one might argue, occurred because of the earlier complacency and lack of adoption of the new technology by domestic producers, which led to a collapse of their market share as the U.S. economy opened up to international competition. More recently, both domestic and world demand for cars and steel has been stagnant, and thus, the productivity gains brought about by new investments to strengthen competitive positions have, at least in the short run, brought about even sharper losses in jobs. To a smaller extent, these problems have also affected the textile and apparel and even the insurance industries. The dynamics in municipal government have turned out to be a bit different, although, in some ways, it also came down to a case of shrinking demand for public services.

In other industries, however, we found that the net employment gains resulting from the expansion of markets have been larger than the employment savings resulting from the higher productivity induced by the new technology. Furthermore, in those industries the new technology itself has often contributed to the expansion of markets by making possible new products or services.

Still, one must observe that even among this second group of fast-growing industries, overall net employment gains may conceal displacement within the firm itself as some activities contract while others expand. Contraction among back-office clerks and among old-styled supervisors and middle managers and expansion among professionals and front-office sales-type personnel are typical.

On the whole, fast growth clearly makes it easier for firms to retrain people for new jobs and thus to avoid laying off people. Nevertheless, we also observed instances of firms that forced employees into early retirement or simply laid them off. In addition, among service industries that rely on large back-office staffs, the reorganization of back offices, which usually involves the consolidation of units and greater decentralization, is often used to bring about labor force restructuring.

It must be noted that among most of the service industries studied, large employers would go to great lengths to avoid up-front firings and layoffs. In the case of low-level clerical personnel, this usually does not pose a major problem because turnover tends to be high and the effects of productivity gains can generally be handled through normal attrition. But problems are often greater among middle managers, who tend to be far less mobile; they are usually older, have reached their positions by moving up through the ranks, and are often limited in their job prospects outside the company. Most firms will try to handle some of these people through a mixture of retraining and early retirement. We found this to be true across a wide span of organizations, from financial service firms, both large and small, to municipal government agencies, universities, and hospitals. Nevertheless, we also saw cases of widespread firing among middle managers, not simply in manufacturing but in some fields of services, such as insurance.

2

Work Force Trends

Introduction

In a 1977 article in *Scientific American*, Eli Ginzberg pointed out that although the U.S. economy had performed well in terms of the number of jobs created since 1950, it had performed poorly in terms of the quality of new jobs.[1] He noted that about two-and-a-half times as many new jobs were added in industries that provided below-average weekly earnings as were added in industries that provided above-average earnings. Yet Daniel Bell, in *The Coming of Post-Industrial Society*, found evidence of a changing economy in which technology was bringing about an upgrading in quality of work,[2] and this latter view is widely held. Is it possible to reconcile these seemingly conflicting views? Where does the truth lie? In which direction are we moving?

In seeking to shed a little light on these questions, let us first look very quickly at the development of the new technology and then at the major transformation that has occurred in terms of the shift to a service economy. We can then examine some trends as reflected in the data, offer some interpretations, and, finally, suggest some emerging problems.

The more than a quarter century of computerization of government and business operations has been marked by a continuous evolution toward wider application of technology in terms of both function and size of the user organization. Although there is as yet no accepted division of these years into stages of development, it is clear that the later years have differed sharply from the earlier ones.

The late 1950s and the 1960s were characterized largely by batch data processing, with large organizations and service bureaus utilizing large computers to carry out high-volume, routine, repetitive tasks. By the end of the 1970s we had entered an era in which complex distributive processing systems were increasingly used by large organizations, and low-cost micro-processing capabilities of one sort or another were applicable across a broad spectrum of functions and to a wide range of firms and public sector organizations.

It is useful when assessing labor market implications to see the mid-1970s as marking a major turning point. At least prima facie evidence is found in the National Income Division estimates of investment expenditures.[3] High-technology investment—if we define this as office and storage machinery, instrument, and photographic and communications equipment—drifted up from slightly more than 0.5 percent of the gross national product (GNP) in 1958 to just more than 1.5 percent in 1976.

From 1976 to early 1982, however, these expenditures rose sharply— to nearly 3.5 percent of the GNP—and accounted for increasingly large shares of all durable equipment expenditures. High-technology equipment accounted for 26 percent of total equipment expenditures in 1976 and 46 percent in 1982.These statistics, however, badly understate the level of investment because software expenditures are much larger. Therefore, what we are seeing is certainly a new era in terms of adoption and diffusion of new technology. The increasing role of the computer has been only a part of a much larger picture.

The Rise of Services

One cannot understand what has been going on in the labor market without recognizing that there has been a major transformation toward a service-oriented economy—a transformation that is quite complex and in which computers may play a role as a catalyst.[4]

The increasing importance of services in our society is not a new development. It goes well back into the nineteenth century, when the rise of the railroads, the telegraph, banking, and wholesales facilitated opening up new agricultural regions and helped usher in the industrial economy. Services have played a far greater role in the transformation of employment in the post–World War II market than previously, rising from 57 percent of all jobs to just over 70 percent in 1985. This figure is somewhat higher than those given by others, but, of course, the statistics depend on what one includes in the category of services. I include utilities, the distributive services, various producer services, government, and consumer and nonprofit services. At any rate, employment by the service industries has risen sharply and these industries provide the lion's share of employment today.

Various sectors of the service economy are also involved. Government employment has increased until very recently, and the nonprofit sectors have made a major contribution. To some extent the consumer services are also involved, although the surprising fact is that they have increased the least of all the services; much more growth has been seen in the business and producer services, including finance, insurance, real estate, and consulting.

This movement toward a service economy has involved a transformation in both what we produce (the change in composition of our final bill of goods in terms of the shares accounted for by government, nonprofit services, consumer services, and manufactured goods) and how we produce (the increased importance of services as an input to production, both those performed in- house and such freestanding services as banking, insurance, consulting, accounting, and legal services). We have gained increasing control over production in plants, reducing employment there. At the same time, we have learned that participating in national and international markets with short product lives and a great need for product differentiation demands a large superstructure of business services in order to carry out development, research, marketing, or distribution.

Changing Composition of the Work Force

Careful analysis of employment data, cross-classified by industry and occupation for the 1960s and 1970s (see Table 2.1), provides a number of insights into how the work force has changed during the past two decades.

The initial observation is simply that employment growth has largely been in the services and has largely been white-collar. Continuous increases in white-collar employment with slow growth or decline among blue-collar workers sharply altered the nature of work during both decades. Major increases occurred in professional, technical, clerical, and service worker occupations.

A second observation is that there have been significant changes in employment in the various occupations that cannot be accounted for simply by industry change. These shifts—increases or decreases over and beyond those accounted for by rates of change in industry employment—provide indirect evidence of changes in various types of work within the economy and are shown in summary form in Table 2.1. In the 1960s, both the services and goods sectors showed positive shifts (disproportionate increases) in clerical, professional, and technical employment. Laborers and service workers, however, declined in relative terms, as did managers and sales workers. The largest positive shifts occurred among professionals, largely due to the increase of teachers.

During the 1970s, the trend toward the increasing importance of technicians and managers characterized virtually every service industry. (This detail is not shown in Table 2.1.) The negative shift in the number of professionals is largely a reflection of retrenchment in the employment

TABLE 2.1
Analysis of Employment Change by Occupation for Service and Nonservice Industries, 1960–1967 and 1970–1976

	1960–1967		1970–1976	
Occupation	Actual Change	Occupational Shift Gain or Loss	Actual Change	Occupational Shift Gain or Loss
Service Industries				
Total	7,486,945	0	8,658,605	0
Professional	2,199,072	401,441	1,179,655	−311,234
Technical	780,785	74,176	1,258,142	504,245
Managerial	467,817	−354,759	1,444,876	483,923
Office clerical	605,205	86,354	967,857	−87,320
Non-office clerical	1,080,542	171,816	828,484	30,868
Sales workers	262,719	−137,059	475,599	−126,106
Craftsmen	457,230	159,925	520,326	153,546
Operatives	436,953	68,527	263,526	−75,433
Service workers	1,094,843	−242,453	1,463,281	−631,501
Laborers	101,779	−55,260	256,859	59,012
Nonservice industries				
Total	1,107,006	0	181,705	0
Professional	225,407	112,772	85,651	82,958
Technical	112,996	29,313	104,123	111,360
Managerial	−1,391,991	−143,063	133,470	97,699
Office clerical	163,548	42,415	−23,651	−22,016
Non-office clerical	197,501	4,258	−57,978	−47,116
Sales workers	31,398	−37,388	7,316	15,268
Craftsmen	503,773	−510	265,452	73,593
Operatives	1,429,580	379,850	−201,419	−118,849
Service workers	−29,850	−74,419	−68,368	−65,087
Laborers	−135,356	−313,238	−62,891	−127,810

Source: U.S. Bureau of Labor Statistics, *Tomorrow's Manpower Needs, National Industry-Occupational Matrix* (Microdata for 1960, 1967, 1970 and 1976) from Thomas M. Stanback, Jr., and Thierry J. Noyelle, *Cities in Transition* (Totowa, N.J.: Rowman & Allanheld, 1982).

of teachers. Although employment gains elsewhere continued to be strong, the demand for professionals appears to have lessened. Among office clericals, the negative shifts were heavy in the distributive and producer services, especially finance, insurance, and real estate, which indicates weakening relative demand in spite of large increases in the employment of office clerical personnel brought on by growth in the service industries. This negative shift among office clericals suggests that new forces—very likely the new office technology—were beginning to be felt. The positive shift in non-office clericals was accounted for principally by increased employment of cashiers in self-service outlets.

The substantial negative shift in the service worker classification is largely because of the sharp decline in domestic servants in the consumer services industrial group. It is probable that to some extent the reduction in the use of service workers occurred because of the application of labor-saving equipment and changing operational procedures.

In the goods sector, there was a substantial positive shift in the employment of professionals, technicians, and managers (executives) but negative shifts for operatives, laborers, and service workers, reflecting a reduction in the demand for production workers and a greater emphasis on administrative and developmental functions. Both office and non-office clericals declined slightly in relative importance, which shows a reversal of the trend of the 1960s toward disproportionate growth and again suggests the influence of the new office technology.

More light is shed on the changing importance of clerical work in a recent study by Matthew Drennan.[5] Drennan analyzed changes in detailed clerical occupational categories for the years 1970–1978, with special attention to six service industries characterized by relatively high levels of white-collar employment. They included banking, insurance, securities, credit agencies, business services, and miscellaneous services. His analysis, which was supported by interviews with managers and executives in important representative firms of these industries, indicated that by the late 1970s computer-oriented technology was beginning to alter the nature of clerical work in many organizations, causing reductions in the need for clerical workers to perform the repetitive and routine tasks.

Detailed data are not available for the most recent years, but such evidence as can be tapped (aggregate occupational estimates for 1979–1981) indicates that the trends begun in the earlier years of the 1970s continue. From 1979 to 1981, the share of total employment accounted for by white-collar workers rose still further, with the growth rate of professionals and technical workers (combined) and of managerial personnel above that of clerical and sales workers. Among the declining blue-collar group, craft worker employment fell relatively less than employment of laborers and operatives. The lower levels of the service worker category continue to expand but at a rate below that of white-collar employment.

Taken as a whole, these analyses reflect clearly the transformations in the service category that have been occurring, while at the same time they support the theory that technology has been reshaping the occupational composition of the work force; that routine work, wherever found, is being curtailed; and that non-routine work is increasing in importance.

Work Force Trends: Good Jobs or Bad Jobs?

Although no final analysis can be offered, it is possible to make several observations that help to put recent trends in perspective. The first observation is that there has indeed been an overall increase in

TABLE 2.2
Distribution of Total U.S. Labor Force Among Earnings Classes, 1960 and
1975, and Distribution of 1960–1975 Job Increases in the Services

Earnings Classes	Distribution of Total U.S. Labor Force (%)[a]				1960–1975 Job Increases in Services[b]		
	1960		1975		Jobs (000)s	Percentage	
1.60 and above	10.9		12.0		1,947	9.5	
		} 31.6		} 34.2			} 35.0
1.59–1.20	20.7		22.2		5,224	25.5	
1.19–0.80	35.9		27.8		2,311	11.3	
0.79–0.40	24.1		28.4		9,205	44.9	
		} 32.5		} 38.0			} 53.8
0.39 and below	8.4		9.6		1,829	8.9	
Total	100.0		100.0		20,516	100.0	

[a] Excludes agriculture, mining, and public administration.
[b] TCU, wholesale, retail, FIRE, corporate services, consumer services, and nonprofit.

Note: These distributions have been computed using 1975 earnings. Hence, the 1960 distribution is a hypothetical distribution showing what the 1960 labor force distribution would have been had workers been paid at 1975 earnings levels.

Source: Based on U.S. Bureau of the Census, *Survey of Income and Education* (for 1975), and U.S. Bureau of Labor Statistics, *Tomorrow's Manpower Needs, National Industry-Occupational Matrix* (for 1960) from Thomas M. Stanback, Jr., and Thierry J. Noyelle, *Cities in Transition* (Totowa, N.J.: Rowman & Allanheld, 1982).

the proportion of low-paying jobs versus high-paying jobs since the 1950s—the position taken by Eli Ginzberg. This shift has come about through the differentially higher growth of services, which on average pay lower wages and salaries, as well as through the declining importance of production workers, who traditionally have earned what we call good American wages. Our analysis of industry occupational subgroups for the period 1960–1975 (see Table 2.2), the most recent years for which we could obtain accurate data, supports this position.[6] Those industry occupation subgroups with annual earnings greater than 120 percent of the national average accounted for 35 percent of all the job increases. Those with earnings of less than 80 percent of the average accounted for 54 percent of job increases during that fifteen-year period. The middle group, with earnings of 80 to 120 percent, was the smallest, accounting for only 11 percent.

Another observation relates to the role of women's work during this period. Between 1950 and 1981, roughly 70 percent of all increases in the total work force can be accounted for by the increase in women's employment. A large share of the rapidly increasing service sector employment has involved work traditionally done by women at low pay. Throughout the period there was a marked tendency for much of the new white-collar work to be defined as women's work, paid for at relatively low rates, and performed by women and, until recently, by young workers from the baby-boom generation.

A third observation has to do with part-time employment. The increasing importance of part-time workers has also been, in large measure, concomitant with the rise of the service sector. Whereas blue-collar work is disciplined by the machine and typically carried out in full-time shifts, white-collar work can often be efficiently organized on a part-time basis.

A final observation relates to sheltering. Typically, workers in the service sector have found less security than those in the goods-producing sector of the sort provided by unions, licensing, or even the work rules and fringe benefits of large organizations. To be sure, many professionals and technicians find protection in their credentials, and some service-producing organizations, particularly public utilities and government, are quite large and have well-established arrangements for seniority and fringe benefits. Nevertheless, for the service sector as a whole, the lack of unionization and the prevalence of small firms, coupled with the greater importance of part-time work, have clearly made for less sheltering.

Taken together, these factors go far to explain why the growth of services has been associated with a major expansion of jobs with relatively low earnings.

Technology in a Changing Economic Environment

In regard to the effects of technological change, both the data analysis and general observations show that a major thrust of the new technology is to root out and eliminate routine repetitive work. Yet, there are contradictory tendencies here for the low-level worker and, in many instances, even for the middle-level worker.

On one hand, the computerized cash register at McDonald's with its pictures of Big Macs and Chicken McNuggets permit the inexperienced young employee to serve the customers very quickly and actually make accurate change. On the other, the word processor creates a need for more skillful and responsible operatives than does a typewriter, and the integrated, interactive, computerized office is likely to require better-trained and better-paid, if fewer, white-collar workers.

It is a puzzle. Where functions are integrated, work often becomes more sophisticated. Where more capital is combined with labor, the worker's responsibility as well as his or her productivity is increased. Yet, one is not necessarily entitled to conclude that an upgrading of work has occurred. User-friendliness makes for simplifications and can make for downgrading. Thus, there are ambiguous tendencies regarding whether work is being upgraded or downgraded.

The important point, however, is that for at least two decades we have witnessed two sets of forces at work. On the one hand, the rise of service activities coupled with the decline of factory work has resulted in increases in white-collar jobs in both upper and lower levels but with the lower-level jobs constituting the large majority. On the other hand, the increasing utilization of a rapidly improving new technology has tended to eliminate routine tasks and to change the nature of work.

Historically, the first set of forces has been in the ascendancy, but the accelerating pace of computerization now promises to give greater scope to the second. This is true not only for the office, the warehouse, and the store but for the factory as well. It includes not simply computers and communications but also other technologies.

My general conclusion is that accelerated adoption of technology is not likely to bring about any radical improvement in the overall quality of employment. Automation or production through Computer Automated Manufacturing or other technologies may cut costs and permit a reduced blue-collar work force to continue to hold jobs in a more internationally competitive manufacturing setting. It may even bring back some manufacturing jobs. It is not likely, however, to be the source of much new employment, nor is it likely to raise industrial wages, given the industrial wage level of other nations. The reality is that we must expect to live in a predominantly white-collar society in which the service sector, with its heavy contingent of relatively low-paying jobs, constitutes the lion's share of employment.

Just how heavily weighted the traditional job distribution of the services is toward low earnings is revealed in the following tabulation for major industry groups. The table shows shares (percentages) of employment earning 80 percent or less of the national average in 1975:[7]

Construction	19.1%
Manufacturing	17.2
Distributive services	9.7
Retail	60.0
Producer services	45.7

Consumer services	82.2
Nonprofit services	48.4
Public administration	6.4
(excluding health and education)	

Only the distributive and public administration categories have a predominance of jobs that pay at an average or above-average level. Among the remaining service sector groups, poorly paid work makes up a very large share (46 percent or more of all employment).

Drennan's data show displacement of work at the lower rungs of the clerical pay scale and some job expansion at levels not too much higher. Work is being upgraded by the new technology in the sense of becoming more sophisticated and entailing more responsibility. Pay scales may be improving as a result, but they are very likely to remain at fairly low levels.

At the technician, managerial, and professional levels, the evidence is not clear. The shift analysis provides tentative evidence that it is in the less well paid technician occupational category that the greatest gains—relative and absolute—are occurring. Moreover, old-fashioned economics should remind us that the user's objective is to decrease, not increase, costs. More and more, the use of the new technology is likely to focus on increasing productivity of higher-level personnel and economizing on their employment.

Emerging Problem Areas

At least two major problem areas related to labor markets appear to be emerging in the mid–1980s. The first relates to inequalities in access to career opportunities among various groups of workers. The second relates to inequalities in opportunities among residents of different geographical areas, especially metropolitan areas.

The more widespread application of the new technology may well create new problems for certain groups in gaining satisfactory work careers. A more rapid diffusion of the new technology carries with it the expectation of an increased demand for employees with computer skills, literacy, and, frequently, a ready familiarity with basic mathematical concepts. Those with superior training or experience in the new technology will enjoy superior employment opportunities; others will not.

For some time it has become increasingly common for large employers to organize recruitment around a two or more track arrangement. New workers with only high school degrees enter the work force only in certain jobs and cannot advance beyond a certain level. Those with college degrees will usually enter in different jobs and will be given an opportunity to advance to higher managerial or professional echelons. Those with advanced degrees may be allowed to move on even faster tracks. Those with no degrees may be denied any track at all. The problems attendant to these new arrangements will be difficult to solve.

A further raising of entry standards is likely to affect the minority youngsters who attend a school where these new skills are poorly taught or who are brought up in an environment in which they have inferior motivation and less opportunity to learn. The record shows that, among minorities, black males have fared the worst in improving their share of white-collar jobs (although they have done relatively well in skilled blue-collar jobs). It seems likely that a more rapid adoption of technology in an essentially white-collar world raises new barriers to equality of employment opportunity and gives rise to an even more urgent need than previously for a careful re-evaluation of our national educational and training policies. The outlook for the factory worker would not appear to be much different. The introduction of highly sophisticated equipment into the manufacturing workplace can hardly be expected to open up job opportunities for unskilled and untrained workers.

The emerging problems associated with variations in employment opportunities among different areas, especially among cities, can only be sketched here. Much has been said in other studies regarding the rise of the "sun belt" areas and the plight of the old industrial centers, but relatively little attention has been paid to how metropolitan areas are being affected by the shift to service activities. Employment data clearly show, however, that factory employment has largely shifted away from metropolitan areas and that where older cities have made successful transformations, it has been through growth in their service sectors.[8] The new successful cities of the sun belt are not specialized in manufactures but in services.

Labor markets tend to differ among regions and among metropolitan areas. Managerial and professional employment, along with employment of large numbers of clerks and service workers, tends to be concentrated in those places that are being favored by the growth of services—those already established as diversified service centers. Cities specializing in goods production tend to have a large proportion of blue-collar workers and a poor service infrastructure. They are, of course, the cities most ravaged by recent waves of industrial unemployment.

One hope for bringing service sector employment to areas of chronic unemployment involves decentralizing back-office types of white-collar work via the use of computerized equipment linked through telephone or satellite communication systems. Thus far, however, there is little evidence that opportunities are opening up for the have-nots of the city system.

To be sure, there has been significant decentralization, but much of it has simply shifted employment to the suburbs. In those instances in which long-range shifts have occurred, corporations have generally seemed to prefer urban centers that are already fairly well endowed with services, usually smaller diversified services centers. There has been no trend toward moving such facilities to places where they are really needed, such as the desolated industrial cities of the Middle West.

Notes

This chapter is a slightly edited version of "Workforce Trends" by Thomas M. Stanback, Jr., a chapter in *The Long-Term Impact of Technology on Employment and Unemployment* (Washington, D.C.: National Academy Press, 1983). Reprinted with the permission of the publisher.

1. Eli Ginzberg, "The Job Problem," *Scientific American*, November 1977, 237:15.

2. Daniel Bell, *The Coming of Post-Industrial Society* (New York: Basic Books, 1973).

3. Goldman Sachs, *The Pocket Chartroom* (New York: Goldman Sachs, February 1983).

4. The discussion in this section is based largely on Thomas M. Stanback, Jr., Peter J. Bearse, Thierry J. Noyelle, and Robert A. Karasek, *Services/The New Economy* (Totowa, N.J.: Rowman & Allanheld, 1981).

5. Matthew P. Drennan, "Implications of Computer and Communications Technology for Less Skilled Employment Opportunities," Conservation of Human Resources, Columbia University, unpublished, 1982.

6. Thomas M. Stanback, Jr. and Thierry J. Noyelle, *Cities in Transition* (Totowa, N.J.: Rowman & Allanheld, 1982).

7. *Ibid.*

8. Thierry J. Noyelle and Thomas M. Stanback, Jr., *Economic Transformation in American Cities* (Totowa, N.J.: Rowman & Allanheld, 1984).

3

Technology, Women, and Work

Introduction: The Changing Role of American Women

My[1] interest in the subject of women and work was first stimulated during my early years of teaching at Columbia's Business School (1937) when I learned from the handful of women students who had returned to the campus for graduate study that even those who had graduated from one of the Seven Sisters colleges with honors had to detour and enroll in a secretarial course at the Finch School or Katherine Gibbs. I was so unsettled by this revelation that I made a special trip to Vassar College to talk with its president, Dr. Henry N. MacCracken, to learn why his graduates were so poorly prepared for the world of work. At the end of a day's explanation I was convinced that with few exceptions, the Vassar faculty was indifferent to the education-work linkage. They assumed that most of their graduates would soon marry and start raising families. The faculty believed that except for the few who were interested in entering professional careers, Vassar graduates could keep usefully busy by actively participating in volunteer endeavors.

Some additional references to these "prehistoric" times are in order. Shortly after the United States entered World War II, I recommended to the Committee on Scientific and Specialized Personnel in the Executive Office of the President of the United States that a roster of women college graduates be started. This recommendation was based on my analysis of the British scene, which demonstrated that after two years of war, women represented the last critical reserve. My proposal was dismissed out of hand. The country was still plagued with high unemployment among men, so high in fact that women who held civil service positions in Massachusetts had to resign if they married!

Twenty years later, in the mid-1950s and early 1960s, the situation was changing with glacial speed. In the mid-1950s, the National Manpower Council decided by only a single vote that "womanpower" was a subject worth exploring. And in the early 1960s, when Barnard College required that its students attend a series of lectures on jobs and careers, my lecture elicited only bored faces and clicking knitting needles. Most of the students were not interested in the advice that I offered: to study calculus and to gain mastery over the quantitative approaches in one of the natural or social sciences, which, I assured them, would provide them not only access to a job, but a job with prospects. Early in the era of the feminine mystique, their minds and emotions were focused in other directions.

These recollections have been presented to make one simple point: The growing importance of women in the U.S. labor market, where they now account for 43 percent of all workers—and this percentage continues to grow—is a recent phenomenon. But during this short time there have been many striking changes in the relations of women to work. Witness the following:

- Over half of all women aged 16 to 64 work, and the proportion is almost two-thirds for those who are at the end of their child-bearing period.
- Although it is true that more women than men work less than full-time, full-year, most women who work, like most men, are regular workers who hold full-time jobs.
- Gender remains a critical determinant of the types of jobs and careers available to women, but it is not nearly as strong a discriminatory influence as it was in the recent and more distant past. In law, medicine, and graduate schools of business, women students account for at least one-third of the graduates, up from less than 10 percent as recently as the mid-1960s.
- For the first time in the nation's history, women outnumber men among students enrolled in colleges and universities.
- In the third of a century since 1950, women have accounted for three out of every five additions to the labor force.
- The explosive growth of the service sector, which today accounted for more than 70 percent of total employment and total output, was both a cause and effect of the availability of women workers.
- Although the anti-discrimination laws and regulations of the 1960s and early 1970s and the changed attitudes and behavior of employers opened up many hitherto restricted fields of work to women (beyond the professions noted above), women continue to be heavily concentrated in a narrow stratum of occupations. Some twenty fields account for two out of every three women workers.
- Over the last half century the occupational group that has experienced the most rapid rate of growth has been clerical workers, which is

a reminder of the need in the present context to consider not only the broad potential impacts of technology but to narrow the focus to specific technologies that are likely to have a strong impact on women workers.

• Again, for the first time in the nation's history, white men no longer constitute the majority of the work force. Women, together with black, Hispanic, and other minority males, today account for more than half of the work force. This last observation suggests that our analysis should pursue a middle road. We must continue to be sensitive to the patterns that underlie the place of women in the world of work, but we must also consider the facts that women account for close to half of the entire work force and their future jobs and careers will therefore be affected by the broad labor market developments that will affect all workers.

There are many other characteristics of women workers that a more elaborate treatment would include, such as the "feminization of poverty," the new sweatshop industries that rely on women immigrants, issues of "comparable worth," the high percentage of women workers in office employment, and still other important facets of women's work profiles. But to contain the analysis, we now shift attention to a limited number of observations about the role of technology and other forces contributing to changes in the labor market.

Observations on Changing Technology

Technological improvements are a way of life for industrial societies, but most innovations involve changes in processes or products that are relatively circumscribed. Even when a significant technological improvement occurs, such as the discovery and manufacture of nylon or the development of the electric typewriter, the impact on the labor market is likely to be absorbed without serious job losses because among other reasons, the lower price or improved quality tends to increase demand. The number of jobs placed at risk by even significant new technology is relatively small and is likely to be stretched out over a period of years. Most mills made the transition from natural to artificial fibers without having to lay off large numbers of workers; the same was true of some of the companies that had earlier manufactured standard typewriters.

There are, however, major technological breakthroughs, such as the development of the railroad, the telephone, electric power, the automobile, and the airplane, where the impacts on work and workers are more pervasive, although these impacts are fully diffused only after long periods, often decades or generations. We will soon celebrate the hundredth anniversary of the first automobile. Some outlying families still are not connected to an electric grid. And in some areas a home telephone is still not affordable by every family.

The microprocessor and the linking of the computer to communications networks give promise of becoming a major technological breakthrough on the order of the development of the railroad and the automobile. It is possible that the computer will prove even more revolutionary because it has the potential of altering not only the movement of people and goods but the nature of work itself (see Chapters 1 and 2 in this volume).

No matter how dramatic the new technology may become, it has been around for a third of a century and it would be difficult to point to its having had large-scale adverse effects on significant groups of workers, male or female, during that period. The most serious charges that can be levied against it is that there have been "silent firings" (that is, women not hired) and other negative effects, such as some deskilling of jobs, downscaling of opportunities, and health hazards. But to date the new technology has been positively correlated with the continued growth of the service sector and with the expansion of women's employment.

When the impact of three sets of forces on the labor market—cyclical, structural (market shifts), and technological—are assessed, the last is surely in the near and middle term (up to one decade) the least important. The Bureau of Labor Statistics has estimated that in the decade of the 1980s only about 15 percent of the changes in the job structure can be ascribed to technology; 85 percent must be sought in the cyclical and structural movements of the economy.

We can now bring this historical perspective to a close by observing that: (1) women's share of total employment has been mounting rapidly and is likely to increase further until women account for half of all workers; (2) this growth has been closely associated with the differentially rapid growth of the service sector; (3) although women are no longer as closely confined to a few major occupational fields, they remain heavily concentrated; (4) the microprocessor and the computer-communications linkages are likely to affect disproportionately the clerical arena in which women workers are heavily concentrated; and (5) even if the new technology were to have a strong impact on existing patterns of work, the consequences would be manifest only over relatively long periods of time.

Policy Perspectives

Framing the Issues

As noted above, the single most important short-term determinant of the labor market experience of all workers will be the growth rate of the U.S. economy and the timing and severity of the next recession. The United States is now in its third year of recovery from two back-to-back recessions, which started in 1979 and ended in 1982. If past is prologue, there is only about one chance in a hundred that if President Reagan remains in office throughout the whole of his second administration, the nation will escape a new recession. As of June 1985, the outlook for the year still appeared to most forecasters to be positive. The expansionary potential does not appear as yet to have spent itself. But it would be an error to overlook the major dislocations that continue to exist: the $900 billion debt of the less developed countries (LDCs); the annual $130 billion deficit of the United States in its foreign trade; the high value of the dollar; the federal budgetary deficits that loom ahead as far as one can see; and the high real rate of interest. If the leading industrial countries fail to attack these five problems conjointly and effectively, the prospect of a severe recession with large-scale labor market consequences is not only possible but probable.

There is steadily accumulating evidence that the United States and other advanced economies are confronting a sea-change in the internationalization of their economies as evidenced by: (1) the spectacular spurt in imports to the United States from the LDCs (with adverse effects in this country on many nondurable manufacturing sectors with large numbers of women workers, such as apparel manufacturing); (2) the relocation overseas of major labor-intensive jobs in durable manufacturing such as low end electronics (again with heavy impact on women workers); (3) the extraordinary transfer of capital funds to the United States, mostly from Europe and Japan, part of which are for investment in plant and equipment and part of which go into money market instruments; and (4) the never-ending trade negotiations that are aimed in the short run at protecting U.S. jobs (as in automobiles and steel) and in the long run to reducing tariff and non-tariff barriers in international trade, with the United States taking the leadership to expand the General Agreement on Trade and Tariffs (GATT) to include services (a high employment area for women workers). As long as the U.S. economy remains expansionary, the odds favor the maintenance and even the reduction in barriers affecting international trade in goods and services. But the United States is no more than one or two steps away from a reversal of its long-term efforts to reduce trade barriers.

And if the next recession is severe or prolonged, it is probable that the nation will not be able to avoid new barriers. Because the United States is the dominant market for the exports of both developed and developing countries, it must exercise restraint to avoid erecting new protective trade barriers. The United States has been through one decade (the 1930s) of beggar-your-neighbor policy. The lessons learned have helped to set and keep the developed world on the path to freer international trade but there is a growing risk that these lessons have begun to fade.

The most important question with regard to the future impact of technology on women's employment is whether the sanguine results of the past twenty years can be projected to the remaining years of this century and even beyond. There are different ways to search for answers. The optimist might contend that there is no reason to expect more "disturbance" in the years ahead than we have experienced over the past two decades, which, as we have noted, saw the computer revolution resulting in few, if any, dysfunctional effects. The pessimist, of course, sees the future differently. In his view, the linkage of the computer to communications networks, which is just now getting into stride, will have a range of adverse consequences for women's employment: first, by eliminating a number of white-collar positions, and second, by making possible the further relocation of many back-office positions from central cities to the suburbs, to distant communities, and also to overseas locations, particularly to English-literate populations such as in Trinidad. The pessimist goes further and points to the downskilling and job and career degrading that often accompany accelerated computerization.

An iconoclast may not be persuaded by either, but would call attention to the following:

1. The computer revolution has reached a point where it is likely to have a greater "displacement" effect on women's employment over the next fifteen years than it has had in the past. A simple projection of the past fails to take into account the fact that computer technology is now on a steeper curve; employers and employees are more willing and able to adapt to it because more than 70 percent of all jobs are in the service sector. Moreover, because U.S. management is striving to become and remain cost-competitive in the world marketplace, it must reduce its white-collar payrolls, and wider reliance on the computer offers reasonable prospects for success in this effort.

2. Analogy is suggestive. Over the past half century employment in agriculture has been reduced from 20 percent to less than 3 percent; in manufacturing, employment has declined from close to two-fifths to less than one-fifth. For a long time neither the tools (computer) and communications resources nor the organizational structure and managerial know-how existed to run large organizations without many layers

of staff. Further, business has learned only recently how to tie large numbers of small units, owned or franchised, into a single organization, but today the United States is the world's leader in these structures—from hotels to fast-food establishments and banks. This experience has proved that the older economists were wrong when they said that services were immune to economies of scale and postulated that they suffered from the "cost disease," with continued dependence on additional labor resulting inevitably in higher costs.

3. But there is more to this than an expectation that the computer will slow and eventually reverse the absolute and relative gains made by clerical workers in past decades. What remains uncertain is whether and how quickly the computer-communications link-up is likely to generate the creation of new products and services that will lead to the employment of large numbers of new workers, both women and men. To engage in an historical experiment, one could identify the types of employment opened up by the widespread introduction of the automobile—from the tens of thousands of people who obtained new jobs in our national parks to millions of construction workers who built homes in the suburbs initially for the wealthy and later for middle-income families. As the time period is extended and the technology becomes more pervasive, it is more difficult and less relevant to assess the impact of a new innovation on total employment. Too many other factors influence the outcome.

Are Women Workers at Risk?

This question is critical for the reasons noted earlier, namely, that such a high proportion of women workers are concentrated within a relatively limited number of occupational groupings, and they account for a differentially larger number of all workers in selected industries. Tables 3.1 and 3.2 provide the critical data.

Several important precipitants can be derived from Table 3.1. First, the thirteen occupational groups shown in the table account for more than half of all women workers. Second, the six occupational fields dominated by women (90 percent or more of all workers) account for one out of every four women workers. Third, what the table does not show is that the occupational distribution of male workers is much less concentrated.

Since our primary concern is to assess the probable impact of technology on women's employment in the remaining years of this century, it may be helpful to consider what happened from 1972 to 1982 to the female-dominated occupational areas. We will single out for special consideration those areas in which technology in general, or

TABLE 3.1
Large Occupational Groups (over 1 million employees) in Which Women
Accounted for at Least Half of All Workers (in millions)

	1982	
	Number	Percent Female
Registered nurses, dieticians, therapists	1.7	92
Teachers, except college	3.3	71
Sales workers, retail	2.5	70
Bookkeepers	2.0	92
Cashiers	1.7	87
Office machine operators	1.1	75
Secretaries	3.9	99
Typists	1.0	97
Assemblers	1.1	54
Food service workers	4.8	66
Health service workers (excl. nurses)	2.0	90
Personal service workers	1.9	77
Private household	1.0	97

Source: U.S. Bureau of Labor Statistics, *Employment and Earnings,* monthly.

TABLE 3.2
Employment by Industry Where Women Accounted for More Than Half of
All Workers in 1970 and 1982

	1970	1982	1970	1982
	(in millions)		(percent women)	
Retail trade	12.3	16.6	46	52
Finance, insurance, and real estate	3.9	6.3	50	57
Personal services				
Private households	1.8	1.3	89	85
Hotels and lodging chains	1.0	1.3	68	66
Professional and related services				
Hospitals	2.8	4.3	77	76
Health services	1.6	3.5	71	76
Teachers, all levels	6.1	7.6	62	66

Source: U.S. Bureau of Labor Statistics, *Employment and Earnings,* monthly.

computer technology specifically, has made significant advances. The total number of sales workers remained stable during the years 1972 to 1982 as did the proportion of women workers. There was a significant increase in the number of bookkeepers, from 1.6 to 2.0 million, and the share of women in the field increased from slightly under 90 percent to slightly over this proportion. The number of cashiers increased at a far higher rate than did bookkeepers, from 1.0 to 1.7 million, but there was no increase in the proportion of women, which remained at 87 percent. Office machine workers expanded from under 700,000 to 1.1 million, and the proportion of women increased from 71 to 75 percent. There was a modest increase in the number of factory assemblers, from 1.022 to 1.087 million, and the share of women increased from 47 to 54 percent.

Several points are worth noting. In a number of fields in which women workers predominated, total employment increased significantly. For the most part, the proportion of women workers as a percentage of all workers did not change appreciably. Most important, the computer and related technology displaced relatively few workers in fields where women were heavily concentrated.

We can supplement our understanding of what transpired in the recent past by looking at employment by industry where women account for half or more of all employees. Table 3.2 underscores that with the single exception of private household employment, which sustained a decline of one-half million, the industries characterized by a predominance of women workers expanded in the years following 1970.

The optimistic implications of this recent experience with respect to both total employment trends and their impact on women workers must not, however, be uncritically projected into the future. It is important to review the U.S. Bureau of Labor Statistics' (BLS) projections to 1995. Table 3.3 reveals that eight out of eleven of the occupational categories for which the Bureau foresees the largest job growth are dominated by women, including secretaries, nurses' aides, sales clerks, cashiers, professional nurses, office clerks, waitresses, and elementary school teachers. Each of these categories will add, according to the BLS, between 425,000 and 745,000 new jobs by the end of 1995.

The forecasts of the BLS in the past have held up reasonably well, at least in total, if not in all subsectors. But the concern here is with selected areas where the new technology is likely to have its greatest impact and where both total employment and the proportion of women workers are substantial. We will look more closely at two industries, banking and hospitals, each of which reveals serious difficulties in assessing the future impacts of technological and other forces on women's employment. The computer and computer-communications linkages,

TABLE 3.3
Occupations with Largest Job Growth, 1982–1995

Occupation	Change in Total Employment (000)	Percent of Total Job Growth
Cashiers	744	2.9
Secretaries	719	2.8
General clerks, office	696	2.7
Salesclerks	685	2.7
Nurses, registered	642	2.5
Waiters and waitresses	562	2.2
Teachers, elementary	511	2.0
Nurses' aides and orderlies	423	1.7

Source: U.S. Department of Labor, Bureau of Labor Statistics, *Employment Projections for 1995*, Bulletin 2197, March 1984.

including satellites, have had a head start in banking and finance, and the new technology has also been making headway, although more slowly, in the administrative, financial, and more recently in the clinical areas of hospitals. Moreover each industry has a large number of workers: banking and finance account for about 2.8 million workers and hospitals employed 4.4 million individuals in 1982, with women accounting respectively for two-thirds and three-quarters of each work force.

One reason that it is difficult to sort out what has been happening in banking and finance is the multiplicity of forces affecting the employment profile. In addition to computers taking over most of the number crunching from clerks, the number of locations where such work is carried out has increased. The new technology is also leading to changes in hiring standards. Most city banks prefer high school and junior college graduates because they have come to recognize that the dynamism of the new technology will require the continuing retraining of staff. The high rate of turnover of new employees, particularly women clerical workers, has enabled most banks to accommodate to the changes to date without layoffs, but they have reduced new hires.

The foregoing is only part of the story. In the past decade, while these changes were occurring in back-office work, many large city banks were opening new branches that required more personnel. Most recently, with deregulation, city banks have moved aggressively to introduce a wide range of new financial services, adding many new workers to fill expanding front-office jobs.

To complicate matters further, the narrowed spread between the rates at which the banks have been able to borrow and to lend, particularly in the early 1980s, caused an adverse effect on their profitability and liquidity. In short, the changes from the side of technology were dwarfed by cyclical and structural alterations that have buffeted and continue to buffet commercial banking. But we must be careful not to minimize the technological factors, as major structural transformations are under way that will permanently transform conventional banks into providers of financial services—of which the full reach remains to be revealed.

A cautionary assessment of the technological impacts on women's employment in banking and financial services would have to include the following:

- A substantial reduction has already taken place in low-skilled clerical positions in insurance and banking as well as in the other sectors of finance, insurance, and real estate (FIRE).
- In addition to back-office clerical positions that have already been relocated out of urban centers, it is likely that additional jobs will migrate to outlying areas and even overseas locations. This trend will have a particularly adverse effect on the employment prospects of urban minority women.
- The raising of hiring requirements will close out most opportunities for young women who do not possess at least a high school diploma and preferably a junior college degree.
- It is likely that many middle management positions will become redundant, which could have adverse effects on many women who have been able to gain a toehold on the executive ladder.
- In contrast to the foregoing, which are "downbeat" forecasts, allowance must be made for the extent to which the new technology will continue to stimulate and possibly accelerate the growth and development of new financial services for which there will be a substantial and sustained demand.

Once this last potentiality is taken into account, there is a reasonable prospect that the long-term employment effects of the new computer-communications technology on women's employment in FIRE will be positive, not negative.

Let us now look at what has been happening to women's employment in hospitals and the changes that are likely to occur in the future. First, hospital employment increased in the twelve years after 1970 by no less than 50 percent, from 2.8 to 4.3 million, and in both 1970 and 1982 women accounted for about three of every four members of the work force. A little-noted phenomenon in this period of expansion in hospital employment has been the trend of most acute care institutions to raise the qualifications of their nursing staff and technicians by hiring and retaining those with more education and training than employees of past years, a preference that reflects the increasing intensity of care and the reliance on sophisticated technology. Many hospitals have shifted their employment patterns in the direction of hiring more registered nurses than before and have reduced the numbers of practical nurses and nurses' aides.

The introduction of the diagnosis-related group (DRG) system for the reimbursement of Medicare patients has acted as a major spur to hospitals to move aggressively to modernize their administrative and financial recordkeeping via computerization. Their survival hinges on how quickly they are able first to understand and then control their admissions, treatment regimens, and length of stay because under DRGs they are paid a fixed price per admission.

As noted earlier, the Bureau of Labor Statistics and most other forecasters assume that health care in general and hospitals in particular will continue to be a major growth industry through 1995. But that assumption must be inspected anew. Hospital admissions have leveled off; length of stay is dropping; the financing of hospital care is being tightened by third-party payers (government and insurance); and the DRG system is encouraging all providers to tighten their cost controls, including their use of personnel. In 1984, total hospital employment, instead of expanding, experienced a small decline, and 1985 shows a further decline.

It would be a serious error to use the past as a guide to the future in forecasting the growth of hospitals because: (1) for-profit enterprises are playing a larger role in the provision of health care than in the past; (2) the future structure of the federal government's financing of Medicare is still evolving, and the numbers of older persons requiring more care continues to increase; (3) medical knowledge and techniques are continuing to advance; and (4) the shift from inpatient to ambulatory settings is accelerating. The error would be especially serious if the focus is on the growth of hospitals, not on the totality of health care services provided.

There is no question that the computer and other new technologies have already left their marks on the hospital indirectly; witness the strong trend towards for-profit and nonprofit chains; the shift from inpatient to ambulatory care settings; radical changes in surgical procedures, particularly cardiac, ophthalmologic and urological surgery; and many other changes, from the introduction of the computer into medical education to the charting procedures used by nurses.

The combined influence of these technological and related changes (economic, organizational, managerial) on the positive employment of women hospital workers are not clearly discernible but the following may be considered a middle-of-the-road assessment.

- The dominant view as of mid–1985 is that hospital employment has peaked and that a decline of up to 20 percent in hospital employment over the next decade and a half is possible; some would say even probable. Clearly, such a decline would have a differentially adverse effect on women workers both because they account for about 70 percent of all hospital employees and because they are more heavily concentrated in the low-skilled occupational categories, which are most vulnerable to the inroads of the new technologies.
- On the basis of selected field investigations in New York City and Boston, my associates and I have become aware of the increasing trend for hospitals to cut back on hiring less-educated and less-skilled persons. This trend means that minority women who improved their employment prospects in the 1960s and 1970s by obtaining jobs in hospitals are definitely at risk. Some are being let go; many more who would have been hired in an earlier period are not even being interviewed.
- It is true that at the upper end of the occupational distribution, women physicians and nurses with a master's degree or a doctorate are well positioned both with respect to employment and advancement. On the other hand, the much-bruited shortage of nurses, which commanded attention only five years ago, has evaporated with little likelihood that a shortage will reappear. Part of this striking shift within such a short time period reflects the increasing pressures on hospital administrators and their ability, through computerization, to exercise much closer control over their personnel costs, particularly their nursing personnel.

For the sake of a balanced overview, one must note that the factors discussed above relate solely to hospital employment, not to total employment in the health care sector. The latter is likely to expand as physicians treat more people in their offices and as more patients, including quite sick patients, are cared for in their homes. Women workers, particularly registered nurses, practical nurses, technicians, and nurse's aides, will unquestionably find that the number of jobs is expanding in these out-of-hospital settings. Total health care employment will probably continue to increase, with a small percentage of women workers—those with high skills—doing better, and many with limited skills having to work at less attractive jobs with little upward mobility.

Directions for Policy

Six major factors have accelerated the growing importance of women workers in the U.S. economy during the post–World War II decades.

- The need for more labor and the availability of women willing and eager to fill such jobs.
- The differentially rapid expansion of the service sector, where employers were often seeking part-time workers, and the fact that many women preferred part-time jobs.
- Many of the new service jobs in clerical work and sales required individuals with a general education and little in the way of specific skills. Women met these basic requirements.
- In the long period of rapid expansion and high profits many corporate employers built up large staffs. Declining profits and the increased use of the computer have in recent years enabled employers to operate with fewer white-collar workers.
- The increasing participation of women in work has been paralleled by a greater percentage of women investing in higher education in order to be able to improve their career prospects.
- Despite these major changes in the relations of women to the world of work, the earlier concentration of women workers in relatively few occupational and industry groupings has continued (although the concentration has been reduced). A high proportion of all women workers continues to be employed in the service sector at the lower end of the wage scale.

In light of these six principal changes in the shape of women's employment, policymakers should be aware of the following:

- More than half of all adult women are currently in the labor force.
- Many more would probably work if suitable jobs were available.
- Women, like men, need employment opportunities if they are to find jobs and enjoy career prospects. The U.S. economy has been slack since 1979 as policymakers have attempted to control inflation. Full employment has been redefined in terms of 7 or 8 percent instead of 3 to 4 percent unemployment. Moreover, the federal government, the only agency capable of affecting macroeconomic policy, has permitted the employment issue to drop off its agenda, although the Humphrey-Hawkins Act obligates both the Congress and the Administration to address a host of job issues. The first and most important contribution of policymakers should be to strive to bring and keep the economy as close to full employment

as possible. At a minimum they should avoid ill-advised actions such as new trade restrictions, radical reforms in the tax structure, and excessively large defense programs that could reduce the capacity of the U.S. economy to move toward a high and sustainable level of employment.

- There is every reason for the government and the corporate sector to maintain and increase their efforts to strengthen their R&D structures, which hold the best promise for the continuing growth and profitability of the U.S. economy in an increasingly competitive world economy. Although new technology has the potential for placing people's skills, jobs, and careers at risk, the penetration of new technology usually proceeds at a rate that permits adjustments to be made through retraining, attrition, and early retirement rather than through job displacement. Although a rapidly penetrating new technology can on occasion result in job losses, most workers who are displaced lose out because of the inability of their employers to remain competitive, as has been the case in steel, autos, apparel, and many branches of electronics.

- Policymakers should recognize that the best approach to the prevention of increasing instability in the world of work is a strengthened educational system that will enable workers to be properly educated and trained and retrained. The economy needs expanded government and corporate funding for retraining programs. The large numbers and high proportion of young minority women, particularly in large urban centers, who fail to graduate from high school need special attention and help. As we have noted, the new technology is leading large employers to raise their hiring standards. Hence young women, including teenage mothers who do not have high school diplomas, may be permanently restricted to the peripheral labor force. These women need more and better second-chance programs such as the Job Corps. The Job Partnership and Training Act (JPTA) is not adequately responsive to the needs of the hard-to-employ.

- The advances in computer-communications technology will, as we have seen, considerably reduce the demand for clerical workers and result in the relocation of many clerical jobs from large high-cost urban centers to outlying and even foreign locations. These developments will make it even more difficult for the urban high school drop-out to fashion a permanent attachment to the labor force, particularly in jobs that offer prospects of advancement. It may be desirable, even necessary, for our society to reappraise the need for a national jobs program (with an educational component) that will help poorly educated young people to acquire work experience and at the same time overcome their educational deficiencies. A national jobs program could also serve as an important bridge for older women, particularly those who have been on and off welfare for some time.

• There is no question that the "crowding phenomenon" referred to above has been a major factor in keeping women's wages considerably below men's and in limiting the opportunities of many women to advance into better-paying jobs and careers. Anti-discrimination legislation and administrative procedures have made some contribution to reducing wage discrimination, but the major positive force has been the expansion of the economy and the willingness and ability of more and more women to prepare for technical and professional careers. Policymakers should continue to use legal and administrative techniques to reduce gross discrimination in the labor market even while recognizing that major improvements in women's earnings and career opportunities depend primarily on the expansionary potential of the economy. Of equal if not greater importance is the response of the urban school systems, which fail to provide many low-income women and men with a proper educational foundation; without this foundation, their entrance into and advancement in the world of work will be seriously circumscribed.

• No one who reflects on women's employment can overlook the importance of strengthening the social service infrastructure, particularly the expansion of child care facilities. Most women who work must also care for their children and run their households. Greater equity and career opportunities for women workers require that society recognize that they carry excessive burdens and seek to lighten these burdens.

A Concluding Note

The foregoing policy recommendations emphasize that the major preconditions for the continued expansion of employment opportunities for women hinge on the continuing strong growth of the economy and on strengthening the educational preparation of women for adulthood and the world of work. Full employment policy and a strengthened educational system are the two principal foundations for further progress. Supplementary support can come from strong anti-discrimination mechanisms and from expanded child care facilities.

But it is unrealistic to expect our economy, or any developed economy, to perform continuously at a high level of employment. Similarly, even a well-functioning educational system will not be responsive to the needs of all young people. A significant minority are likely to reach working age inadequately prepared for the world of work. Large-scale shifts in markets and new technological breakthroughs introduce further disturbances that will result in job losses, skill downgrading, and reduced earnings even while they also open up new opportunities for job growth, skill improvements, and higher earnings. A responsible and responsive democracy must act to assist those who are most vulnerable to the malfunctioning of the schools and the labor market. It can do so by providing second-chance opportunities for the many unemployed persons who need to improve their basic competences if they are to be successful in obtaining a private sector job; interim public employment if they are not capable of competing successfully for such jobs; and access to training and retraining in the event that they are victimized by market or technological change.

Our society confronts a paradox that it can no longer ignore. It cannot hold onto its conviction that all persons should work to support themselves and their dependents and at the same time ignore the reality that many people lack the required competences for getting and holding jobs and that many others, even those who are competent, cannot find jobs. If the United States reaffirms its commitment to the work ethic, it must see that everybody, men and women alike, who needs or wants to work has an opportunity to do so. Alternatively, society can turn its back on the work ethic and recognize that it must support large and growing numbers of individuals who are unable to support themselves. But if we choose the latter alternative, we run the risk that, divided between those who work and those who are cut off from work, our society will not long endure, surely not as most of us would like to see it endure.

Notes

1. This chapter is a slightly edited version of a paper by Eli Ginzberg, "Technology, Women and Work: Policy Perspectives," presented in 1985 for the Panel on Technology and Women's Employment of the National Academy of Science.

4

The New Technology and
Equal Employment Opportunity

Introduction

Are sex and race still major determinants of employment discrimination? Is the new computer technology changing the demand for labor to such an extent as to reshape the terms under which women and minorities encounter discrimination? As I argue in this chapter, both questions deserve qualified answers.

Although major advances have been made in alleviating sex and race discrimination in the workplace, such discriminatory barriers remain. But other discriminating characteristics that played a lesser role in the past loom larger today. For example, there may be an increasing tendency to use age to differentiate youth and some groups of older workers from others in the labor market. The age factor seemed relatively unimportant during the 1950s and 1960s. More important, socioeconomic status may be growing in importance because of its implications for access to formal schooling and higher education. Although our economy places increasing value on formal training as a criterion for hiring, our society continues to lag in providing equal access to quality education.

The new technology is clearly changing the nature of many jobs and skills and, as a result, redefining the demand for labor. In the process, it is loosening the hold of particular groups of workers on specific occupations; this loosening, however, need not be restricted to women and minority workers. At the same time, the new technology cannot be viewed in isolation. The new wave of technological change often simply reinforces developments inherent in years of deep economic and social changes during which industry shifts acted to alter the overall occupational structure and formal schooling and higher education became increasingly important in determining who gets access to what job.

The overall thesis of this chapter is straightforward. The current labor market transformation is fundamentally altering employment and mobility opportunities, altering in particular the need for training and the way training is provided, and in the end altering an earlier balance in the sources of discrimination.

A principal conclusion of the chapter is that not only must direct Equal Employment Opportunity (EEO) enforcement in the workplace continue, but the reach of EEO enforcement must be widened—in part by broadening an earlier, almost exclusive focus on the workplace to one that links the workplace to the educational arena. At the federal level, this shift will require an expansion in the scope of activities of the EEO-Commission, perhaps even new legislation. Under Title VII of the Civil Rights Act of 1964, the scope of the Commission is largely restricted to the workplace.

The two main parts of this chapter develop the argument that leads to this conclusion. The first major part emphasizes key changes that have taken place on the demand side of the labor market; the second, adjustments that have occurred on the supply side, with special references to various groups of women and minorities.

Changes in Labor Demand

The Rise of the New Service Economy

The U.S. economy has been, for some time now, in the midst of a major transformation that involves the shift of capital and labor out of the smokestack industries and into high-tech and service industries. Although this transformation has been in the making through much of the postwar period, the 1970s represent a watershed. The formidable acceleration of the internationalization of the economy after the first oil crisis of 1973 contributed greatly to speeding the redeployment of resources, as many older industries were put through the wrenching test of worldwide competition.[1]

Between 1970 and the last quarter of 1984, a net 27.2 million new jobs were added to the economy, of which nearly 95 percent were in the service industries.[2] Looking at the Reagan years only, the shift to services was even sharper. By November 1984, employment in the goods-producing industries—agriculture, mining, construction, and manufacturing—had not even caught up with their January 1981 level.[3] In net terms, these statistics mean that employment growth since early 1981 had been 100 percent in the services.

In occupational terms, the labor market transformation of the past two decades has translated into more than seven out of every ten workers being employed in either white-collar or service worker occupations. This proportion has been the result of a very sharp drop in the share of blue-collar workers—from 39.2 percent to 29.3 percent of the non-agricultural labor force between 1965 and 1983.[4] Growth has shifted to service industries dominated by white-collar or service worker occupations and technological changes in manufacturing have accelerated the shift to managerial, engineering, technical, sales, and clerical occupations.

In broad terms, these shifts have been tilted toward women and minority workers. During the same 1970 to 1984 period, nearly two-thirds of the new jobs have been filled by members of these groups. By late 1984 white males for the first time no longer constituted the majority of the labor force; their share of employment had dropped from nearly 55 percent in 1970 to 49.5 percent by late 1984.

The Early Years of EEO:
Opening Internal Labor Markets
to Women and Minority Workers

The aforementioned statistics mean little until one analyzes who gets hired for which jobs and through which mechanisms. The record of the early postwar period, roughly speaking until the mid- and late 1960s, reveals the surprising extent to which employers used to rely on internal labor market structures to prepare and move workers to staff most of the ranks of their organization and the rapidity with which this practice has changed since then.[5]

The reliance on "internal labor markets" was extensive not only among the manufacturing giants that typified the era—the IBMs and the GMs—but also among many medium-sized firms, including those in the service sectors.[6] This factor is important to stress. The recent loosening of the reliance on internal labor markets cannot be ascribed solely to the overall industry shift to the services, but must be seen as part of a total labor market transformation affecting both manufacturing and service industries.

In the insurance industry, for example, most workers in the past entered the job ladder straight out of high school, starting at the very bottom of the organization as messengers or file clerks. Through on-the-job training and seniority, they would move up the ranks. The most successful would move gradually from an entry-level clerical level into a professional position from, say, messenger to statistical clerk, claims examiner, or policy rater and later to assistant underwriter or underwriter.[7] In the department store industry, workers would enter as stockroom clerks, move into sales positions, possibly to a commissioned sales position in high-ticket departments (furniture, household, appliances, etc.), and even to department manager, assistant buyer, and buyer positions.[8]

Important differences could be found from industry to industry and from firm to firm. In the construction sector, for example, mobility ladders were industry- and craft-based rather than firm-based, with trade unions often playing a central role in operating the mobility system.[9] In addition, most small firms lacked both the resources and the range of employment opportunities necessary to operate internal labor markets and relied extensively on the open labor market. More important, perhaps, sex and race stereotyping was often used to create sex- or race-labelled occupations. In turn stereotypes were used to restrict mobility opportunities available through internal labor markets to white males, by channeling women and minority workers into dead-end jobs. A good deal of the mid–1970s literature on internal labor markets sought to account for all of these differences and the way they contributed to discrimination toward different groups of workers.

Consistent with the dynamics of labor markets prevailing at the time, a principal focus of EEO policy, when it was first formulated, was to open internal labor markets—through both hiring targets and internal targets—to those who for reasons of race or sex had been left out or left behind. Much attention was focused on industries that were then the pillars of the economy: manufacturing and the public sector. The efforts of the federal government to accelerate the promotion of minorities and women within its own agencies as well as efforts within the private sector—through major consent decrees such as those secured in 1973 between AT&T and the EEO-Commission[10] and in 1974 between the Commission and the steel industry[11]—typified that period of EEO enforcement.

Substantial gains were achieved during the 1970s through these decrees. In retrospect, however, we can see that these efforts were focused on industries and work settings that were declining in importance in the economy and that they offered only limited clues as to how to approach and solve labor market discrimination in today's economy. Over the past decade or so, the role of internal labor markets has weakened dramatically across a broad range of industries. The reasons are numerous and diverse, but they add up to the growing pressure on firms to externalize the cost and responsibility for the training process, to rely more extensively on external labor markets for new workers, and to put in place new arrangements affecting who they hire and promote.[12] Two primary forces are responsible for this new dynamic: the substantial expansion of schooling and higher education throughout the postwar decades and the new wave of technological change.

The Impact of Education on Hiring Requirements and Mobility Ladders

The first factor behind the transformation in hiring and mobility opportunities—the postwar expansion of the schooling and higher educational system—was slow in the making but nevertheless irreversible. It was largely a case of supply changes leading to demand changes. By changing the makeup of the labor supply, the expansion of the educational system put pressure on all firms to adjust their hiring procedures to the new availabilities of a labor supply increasingly differentiated by grades and types of education. For example, whereas only slightly more than 10 percent of those between ages 25 and 29 had received four years or more of college education in 1960, by 1980 their share had risen to nearly a fourth.

The expansion of formal education led to a major shift to outside hiring, first felt most strongly at the level of professional and managerial personnel—the so-called "exempt workers." One consequence was the weakening of some traditional internal ladders, especially those designed to move the ablest workers from nonexempt positions into supervisory and middle managerial positions. No longer could a sales clerk expect to become a buyer for a major retailing organization or a messenger expect to become an insurance executive by simply moving through the ranks. Rather, most companies began recruiting exempt workers directly from college.[13] In that respect, the 1970s represent somewhat of a turning point as the cumulative effect of several decades of expansion of the educational systems and the coming of age of the baby boomers were finally being felt massively on the supply side of the labor market.

The Impact of the New Technology on Skill Requirements

Whereas earlier changes in hiring and mobility opportunities had been mostly supply driven, recent changes have been largely demand driven. They are the result of the introduction of the new computer-communications technology and its impact on skills. Broadly speaking, the new technology has acted to reinforce the tendency toward a weakening of internal ladders. Two preliminary observations are warranted to establish this point.

First, vast areas of work are being transformed and reorganized around the processing of information through interaction with computerized systems. Thus far, the areas most directly affected lie primarily in the middle range of occupations, from relatively low-level clerical positions or even blue-collar operative positions all the way to low-level or middle-level professional workers.[14] Still, high-level technical, professional, and managerial work will not long remain unaffected. Only in the case of the lowest level occupations—primarily among laborers, service workers, and low-level sales and clerical classifications—has the new technology, thus far, had little or no direct impact on work and skills. It may be relevant to note here that these lowest occupations, including sales clerks, cashiers, building janitors, guards, orderlies, cooks, and others have been among the fast-growing areas of employment[15] and that they are conspicuous by their absence of mobility ladders.

Second, the new technology does not, as many initially believed, lead ineluctably to downskilling but rather to varying degrees of upskilling. This generalization does not preclude occasional downskilling or occasional lags between current and potential uses of technology by firms. Upskilling comes about for three principal reasons: first, the most efficient use of the new technology often seems to lead to a reintegration of tasks previously parcelled out among different workers; second, as intelligent systems take over "processing functions," workers are left with "diagnosis" and "problem-solving" functions; and third, the shift to "problem-solving" functions at the low levels of the organization calls for a simultaneous decentralization in decisionmaking power.[16]

As the new technology changes the skill requirements for many jobs, it also leads to the homogenization of skills across a wide range of industries, encouraging the externalization of training for many middle-level workers. For example, the job of bank clerks processing letters of credit or fund transfers on a computerized system, of insurance examiners processing claims, of airline agents processing reservations and ticketing, or even of telephone switchmen routing and managing traffic flows through switches are becoming not only more demanding in terms of skills, but also increasingly similar in terms of skills required.[17] Not surprisingly, a major focus of the current "training debate" about the need for more sophisticated training institutions is on this middle range of occupations, because these are occupations that in their older configurations had rarely been brought within the purview of formalized training processes. These were jobs for which skill training was traditionally acquired on-the-job through internal labor market mechanisms. Hirshhorn refers to this transformation as the process of "para-professionalization."[18] Thus, the institutions most directly concerned by the new demand for training are clearly not simply the high school, but even more so the vocational-educational institutions, the community colleges, and even the four-year colleges.

Technological Change and Increasing
Institutional and Geographical Mobility

Because formal education and training have become increasingly important in determining a worker's position in the labor market, there is a presumption that better-prepared workers should have an edge, particularly in terms of improving their earnings. Thus far, this has not necessarily been the case. The tendency toward universalization/homogenization of skills has also weakened the degree to which workers are sheltered from competition as they once were when skills were more specific to the industry or the firm. Further, this has been aggravated by a context of weakening unionization.

In addition, the new technology makes it increasingly feasible and cost efficient to separate geographical so-called back-office functions (dominated by clerical and service worker occupations) from front-office functions (dominated by technical, sales, professional, or managerial occupations). Two consequences follow. First, the separation contributes to breaking the institutional job linkages that used to exist when entire departments, from bottom up, were located in the same physical location. Second, the increasing mobility of back-office establishments puts workers on the defensive because geographical mobility is difficult for the increasing numbers of two–wage earner households.

Adjustments on the Supply Side:
The Shifting Nature of Discrimination

The broad changes that have taken place on the demand side of the labor market as well as the gains made by certain groups of workers in selected occupations and industries as a result of early EEO efforts have acted to shift the nature of discrimination.

Two examples will serve to illustrate some of the shifts. The first one covers back-office clerical employment; the second one, sales employment in the retailing sector. In the first case, technology is brought in to directly reform, reorganize, and rationalize work in areas with large concentrations of workers. As suggested earlier, the impact is often some degree of upskilling and associated changes in the demand for labor. In the second case, the direct impact of technology on sales and related occupations is relatively modest. The impact of technology is mostly indirect in that it permits great improvements in the control and coordination of the organization itself (in buying, inventory control, and accounting). To the extent that changes in labor demand can be observed, they are unlikely to be associated directly with technology.

These two examples cover major work situations in which great numbers of women and minority workers have traditionally found and continue to find employment.

The Reorganization of Back-Office Employment

This first example relates to the reorganization of clerical work typical of the back offices of banks, insurance companies, telephone companies and other utilities, and other organizations with large processing facilities. Back in the 1960s and early 1970s, these firms hired large numbers of youth as messengers and file clerks straight out of high school to staff entry-level clerical positions. Later, these young workers would be trained in-house and would move up the ladder as they matured.

As Appelbaum has noted, the long-standing tendency in back offices was to discriminate between white men and women and minority workers by operating a two-track system.[19] One track, reserved mostly for white men, could gradually lead those men into professional or managerial employment; the other, used primarily for women and minority workers, would channel most of them into dead-end positions. By forcing companies to do away with these practices, EEO, for a time at least, opened new avenues of opportunities to women and minority workers. Yet, no sooner had these avenues been opened than their reach was considerably curtailed as the result of the tendency toward the weakening of internal labor markets and, in particular, the delinking of nonexempt from exempt jobs. This trend did not completely shut out access for women and minorities to many managerial and professional positions; but, typically, it forced them to enter through another route, that of higher education. At the same time, changes were also occurring at the traditional entrance level, with impacts on both adult workers of different races and on youth.

For many years companies with large back-office employment were known for their close links to the local high schools. During the past decade, however, this situation changed dramatically as a result of the new technology. As one executive of a large New York insurance firm reported in a recent interview: "Up until the early 1970s, we hired nearly 2,000 kids every summer. Today, we hire at most 100 kids. Nowadays, most entry takes place at a higher level—typically community college or equivalent—straight into claim examiner positions. Most of the filing and messenger functions have been eliminated through computerization."

In a recent study of the youth labor market in New York City, Bailey and Waldinger found that of the nearly 40,000 jobs lost by youth in New York City during the past decade (1970–1980), nearly half of those losses could be attributed to the sheer contraction of the city's economy and the other half—that is, nearly 20,000 jobs—could be attributed to the elimination of filing clerks, messenger clerks, and similar positions in local public utility companies (telephone, gas, and electric), in banks, and in insurance firms.[20] Beyond the magnitude of these numbers, these losses implied that by the late 1970s a major group of workers—youth with high school or equivalent diplomas—no longer had available to them entry-level opportunities with built-in promotion ladders. Instead, they had largely been relegated to entering retail and consumer services, which typically offer far more limited opportunities for upward mobility.

The trend just discussed, which was set in motion in the early 1970s when large back-office organizations began investing in centralized electronic data processing and eliminating many paper files, is being followed by yet another trend, which is growing out of the deployment of distributed data processing in the late 1970s and early 1980s. The new generation of computer-communications technology permits a geographic separation of these back offices from front and head offices of the firm. It permits the parent organization to seek new locations away from the central districts of very large cities such as New York, Los Angeles, Chicago, Philadelphia, and other places where they have traditionally been located. The greatest impact of this new trend appears to be on minority women, who had made great gains in entering the clerical ranks during the 1970s but may now be left behind in the inner cities where they reside while back-office jobs are being moved elsewhere.[21]

Although some groups are losing, others are clearly gaining from this restructuring/relocation of back-office work. In general, employers relocate their facilities not only in areas where operating costs (rent, utilities) and labor costs are low, but also often in areas where they can find an infrastructure of community colleges (or equivalent) that will help them prepare and train employees. Typically such moves bring firms to the suburbs where they seek large pools of middle-aged, married, usually white women. In some cases, it brings employers to communities with large military installations where they hire both married wives of enlisted men and retired military clerks willing to put in a few more years of work. The advantage of hiring from these groups is that these are workers who typically demand little by way of mobility opportunities, something that most employers can no longer offer because of delinking between back-office clerical positions and higher level jobs.

The Transformation of Retailing Employment

Retailing is one sector where employment transformations appear to be less directly linked to recent or past technological changes. This does not mean, however, that the new technology has not found its way into this sector, but that its impact has been more diffuse and indirect than in back- office employment.

The postwar period witnessed the rapid growth and diffusion of large chain organizations. Until the 1940s, organizations such as Sears and A&P were exceptions. The postwar period saw a rapid growth of multi-unit organizations in foods, dry goods, hardware, gasoline, and many other areas penetrating markets traditionally dominated by "mom-and-pop" businesses. The resulting shift in the scale of operations made possible substantial rationalization, accompanied by major productivity gains in buying, inventory control, and accounting, which facilitated the further growth of large sales organizations with relatively small administrative staffs. These changes also made it easier for large retail organizations to follow their customers into the suburbs, where they were able to tap into underutilized pools of suburban married women who were often eager to work.

Simultaneously, the cumulative effect of changes in work habits and spending patterns (e.g., two-worker families) led retail organizations to stay open for more hours during the week and to make more use of part-time employees. In the place of the basic 9-to-5, 40-hour workweek, many retail organizations today are open 65 hours a week (10 hours on weekdays; 7.5 hours on Saturday and Sunday). In some of the largest metropolitan areas, supermarkets compete on a 24-hour, 7-day-a-week basis. The impact of these changes on employment patterns has been dramatic: The reliance on part-time employees has skyrocketed. In department stores, the breakdown between full-time and part-time employment shifted from 65 percent full-time versus 35 percent part-time in the mid–1960s to the reverse ratio nowadays as stores have added half-time and short-hour shifts.[22]

A great deal of the employment expansion in retail was first based on the hiring of women—first white women, later minority women. Clearly, the expansion of part-time jobs was aimed in part at facilitating the employment of married women, many of whom preferred not to put in a full workweek. Large retailing organizations were leaders in uncovering those underutilized pools of women, but the discovery did not go unnoticed for long in other industries. As noted previously, large clerical organizations are now actively seeking such employable women as they relocate back-office facilities to the suburban rings of large cities. In the process, these organizations are creating new pressures on the female labor market. As a result, many retail organizations are now seeking to recruit more actively from among high school youth, many of whom are now available because, short of educational credentials higher than a high school diploma, they are blocked from competing for more desirable jobs.

A Look at the Aggregate Data

What can be discerned about the critical linkages between technology and discrimination from employment aggregates? To seek some answers, I developed two sets of measures—one based on decennial census data, the other on EEO-Commission data.

Table 4.1, based on 1970 and 1980 census data, presents an industry shift-share analysis for six major groups of workers: youth (16–19), black females, Hispanic females, white females, black males, and Hispanic males. For each group, employment growth (or decline) in an industry has been broken down among three components: first, growth (or decline) associated with the relative growth (or decline) of the industry; second, growth (or decline) associated with an increase (or decrease) in the group's participation in the employed labor force; and third, growth (or decline) associated with the pure shift of the group in or out of the industry. In other words, the shift measure indicates the gains or losses in a given group's penetration in a particular industry, everything else being held constant. For each group, the shift measure is shown in terms of a "turnover," that is, the number of workers "shifted around" during the ten-year period shown as a percentage of the group's 1980 employment. In addition, the positive and negative shifts are distributed in percentage terms among industries. The overall impact of the shift is shown by means of normalized shares for 1970 and 1980 that show the group's relative standing in an industry vis-à-vis the total labor force. A ratio below 1 means that the group penetration in the industry is lagging; a ratio above 1 means that the group is overrepresented.

For example, Table 4.1 shows a turnover measure of 2.6 percent for white women, indicating that in 1980 2.6 percent of the 34,806,839 white women ended up in different industries from the ones in which they would have been employed had there been no change in the penetration of white women in various sectors of the economy between 1970 and 1980. In addition, the table shows that this shift was due primarily to increased penetration in public administration (explaining 25.8 percent of the positive shift), other goods (+20.9 percent), and FIRE and business services (respectively +14.7 percent and +11.5 percent), matched by decreased penetration in education (explaining 37.9 percent of the negative shift), manufacturing (−32.1 percent), and health (−28.4 percent).

On the whole, Table 4.1 points to the following: Shifts were extensive among black females (turnover of 13.9 percent for the entire group) and youth (8.3 percent turnover) and rather limited among white females (2.6 percent turnover), Hispanic males (3.3 percent), and black males (4.3 percent).

The greatest exit-move of black females was out of the consumer services industries where large numbers were once employed as domestic servants. Their greatest gains were in manufacturing (+29.4 percent); public administration (+18.2 percent); FIRE and transportation, communications, and utilities (TCU) (+29.1 percent combined), where they made substantial gains in clerical work; and in the educational sector

TABLE 4.1
Industry Shifts of Major Groups of Workers and Distribution of Positive and Negative Shifts Among Industries

	Employment Distribution All Sex, Race & Age		Youth (16–19 Yrs. Old)				Black Female				Hispanic Female			
	1980 %	1970 %	Normalized Share 1980	1970	Shift (%) +	–	Normalized Share 1980	1970	Shift (%) +	–	Normalized Share 1980	1970	Shift (%) +	–
1. Health	7.4	5.5	0.68	0.93		–21.3	2.35	2.35	+ 0.3		1.42	1.81		–23.6
2. FIRE	6.0	5.0	0.67	0.85		–13.2	1.03	0.70	+15.1		1.23	1.17	+ 3.4	
3. Social	1.8	1.6	0.87	0.70	+ 3.1		2.13	1.32	+10.5		1.48	1.18	+10.7	
4. Bus. Serv.a	6.6	5.7	0.72	0.90		–14.7	0.60	0.85		–10.6	0.73	0.83		– 6.6
5. Education	8.6	8.0	0.62	0.75		–13.5	1.59	1.49	+ 9.4		1.14	1.11	+12.4	
6. TCU	7.3	6.8	0.33	0.59		–23.2	0.71	0.44	+14.8		0.52	0.49	+ 6.8	
7. Wholesale	4.3	4.1	0.71	0.68	+ 0.8		0.34	0.32	+ 0.8		0.71	0.81		– 7.3
8. Construction	5.9	6.0	0.71	0.52	+12.9		0.10	0.06	+ 1.4		0.14	0.11	+ 0.8	
9. Public Adm.	5.3	5.5	0.40	0.37	+ 1.9		1.51	1.07	+18.2		0.90	0.73	+16.8	
10. Cons. Serv.b	20.3	21.4	2.46	2.11	+80.6		1.06	1.70		–86.1	1.23	1.39		–62.5
11. Other Goodsc	4.0	4.5	1.06	1.02	+ 0.7		0.20	0.33		– 3.4	0.59	0.56	+ 0.2	
12. Manufacturing	22.4	25.9	0.60	0.64		–14.1	0.78	0.60	+29.4		1.14	0.96	+52.3	
	100.0	100.0			100.0	100.0			100.0	100.0			100.0	100.0
1980 Employment			6,973,441				4,659,177				2,168,649			
Turnover (% 1980 empl.)			8.3%				13.9%				5.7%			

a legal, accounting, advertising, and the like
b retailing and personal services
c agriculture, mining

TABLE 4.1 continued

	White Female				Black Male				Hispanic Male			
	Normalized Share		Shift (%)		Normalized Share		Shift (%)		Normalized Share		Shift (%)	
	1980	1970	+	−	1980	1970	+	−	1980	1970	+	−
1. Health	1.74	1.93		−28.4	0.65	0.58	+ 7.7		0.42	0.46		− 3.4
2. FIRE	1.43	1.41	+14.7		0.58	0.53	+ 4.5		0.54	0.61		− 8.3
3. Social	1.45	1.32	+ 8.1		0.84	0.86		− 2.1	0.48	0.61		− 7.4
4. Bus. Serv.[a]	0.92	0.90	+11.5		0.82	0.84		− 4.9	0.98	0.96	+ 7.2	
5. Education	1.53	1.70		−37.9	0.68	0.54	+16.4		0.42	0.45		− 6.9
6. TCU	0.56	0.59		− 1.5	1.72	1.45	+38.5		1.12	1.17		−11.3
7. Wholesale	0.67	0.65	+ 4.7		0.96	1.00		− 7.2	1.18	1.25		−11.3
8. Construction	0.21	0.17	+10.8		1.37	1.56		−29.5	1.70	1.47	+34.4	
9. Public Adm.	0.88	0.77	+25.8		1.33	1.32		− 1.7	0.87	1.12		−44.0
10. Cons. Serv.[b]	1.30	1.28	+ 3.4		0.72	0.74		−16.7	0.90	0.90		− 2.9
11. Other Goods[c]	0.40	0.27	+20.9		0.88	1.30		−37.8	2.04	1.97		− 4.5
12. Manufacturing	0.72	0.77		−32.1	1.29	1.21	+32.9		1.26	1.10	+58.4	
1980 Employment	34,806,839	100.0	100.0		4,674,871	100.0	100.0		3,288,208	100.0	100.0	
Turnover (% 1980 empl.)	2.6%				4.3%				3.3%			

Note: The twelve industries are ranked by rate of growth between 1970 and 1980 from the fastest growing (health) to the slowest growing (manufacturing). The industry breakdown used is based on the classification of service industries developed in Thomas M. Stanback, Jr., Peter J. Bearse, Thierry J. Noyelle and Robert A. Karasek, *Services/The New Economy* (Totowa, N.J.: Rowman & Allanheld, 1981). The first two columns show the distribution of all employed among the twelve industries in 1970 and 1980. The positive and negative "shift" is distributed for each group on a percentage basis. The normalized shares of major groups of workers shown for 1970 and 1980 are computed by dividing the share of employment held by each group in each industry by that same group's share of all employment in all industries combined. An index below 1.00 indicates underrepresentation; above 1, overrepresentation. For definition of "shift," see text.

Source: U.S. Census of Population, *Detailed Characteristics of the Labor Force*, 1970 and 1980.

(+9.4 percent). Youth's greatest losses were in TCU, FIRE, and business services (51.1 percent of their losses combined), and their greatest gains were in retailing (a staggering 80.6 percent). These findings are highly consistent with the analysis presented earlier.

As for other groups, the patterns of gains and losses among Hispanic females tended, with some discrepancies, to resemble those of black females. Among black males and Hispanic males, the common findings are that their limited positive shifts were overwhelmingly concentrated in some of the least dynamic and slowest growing sectors of the economy: manufacturing and TCU for black males (+71.4 percent), and manufacturing and construction for Hispanic males (+92.8 percent). Last, the outstanding finding for white women remains that industry shifts over the decade were very limited, with an overall turnover of only 2.6 percent.

The second set of data, presented in Table 4.2, shows changes in the normalized shares of major groups of workers in major occupations between 1966, 1978, and 1981. These data are only for large private-sector firms (100 employees) and are based on EEO-1 reports, which large employers must file every year. They show the progress made by various groups of workers in what has been traditionally the most progressive sector of the economy in terms of EEO enforcement—large employers.

Some major changes are worth highlighting. The data show white women advancing out of clerical positions and making large gains in professional ranks; black women and Hispanic women moving out of service worker and laborer positions, respectively, and, in both cases, gaining in clerical positions; and black men and Hispanic men shifting out of laborer positions and into operative and craft positions. Despite some scattered gains, minority males in general continue to trail considerably behind in the fast- growing white-collar occupations. Of all groups, Hispanic males appear to be the most immobile.

These data are consistent with several, if not all, of the trends suggested earlier in this chapter. Clearest among the trends identified are: (1) a relative narrowing of job opportunities for youth; (2) a relative improvement in the position of women as shown by the gains of minority women into clerical positions and the advances of white women into professional positions; and (3) a general lack of progress by minority men in entering the relatively fast-growing service industries and white-collar occupations.

In concluding this section, a note must be added about the tendencies toward age-based discrimination. Aside from the patterns that have emerged among youth, the kind of statistical analysis developed in this section appeared too crude to yield significant evidence of age-based discrimination among other groups. This, however, does not rule out other such forms of discrimination.

Conclusion

For over twenty years now, the United States has had a policy of equal employment opportunity enacted into law, administered by a specialized federal agency, and enforced through the courts.

Because EEO was shaped under specific historical circumstances—namely, as an outgrowth of the civil rights movement and, later, the women's movement—and in response to the reality of the labor markets of the 1960s and early 1970s, the principal emphasis of early EEO policy was the elimination of sex-based and race-based discrimination in the workplace. Workplace discrimination, at the time, was primarily rooted in blatant sex- or race-job stereotyping. It was perpetuated not simply through cultural biases but quite concretely by excluding women and minority workers from entering jobs traditionally held by white men and from accessing opportunity ladders available to white men. In an environment in which the majority of workers were rarely educated beyond high school, formal education was seen as playing a relatively minor role in determining what happened once workers had entered the labor market, except for a few college-educated individuals.

To assert that sex or race discrimination in the workplace has been eliminated and no longer needs the nation's attention would be both silly and wrong. But it would be equally wrong to write off the past twenty years of EEO enforcement and to assert that nothing has changed.

In this chapter I have suggested that what is needed is: (1) an assessment of the changes that have occurred in the labor market partly as a result of earlier EEO efforts, partly as a result of the rising importance of education, partly as a result of technological change, and partly as a result of the structural shift from manufacturing to services; (2) an assessment of the impact of these changes and of their role in bringing to the fore factors of discrimination other than sex or race; and (3) an assessment of the way these new factors of discrimination may be used either independently or in relationship with gender or racial characteristics to bring about different patterns of discrimination. One such change emphasized in this chapter is the rising importance of formal education in determining labor market outcomes.

Formal education is clearly becoming a major determinant of a worker's long-term position in the labor market. This trend is not simply a case of growing credentialism for the sake of erecting new barriers, even though the tendency may be at work. Professions have traditionally used formal accreditation or licensing based on educational degrees as

TABLE 4.2
Employment of Major Groups of Workers by Sex, Race, and Occupation in EEO Reporting Firms:
All Industries, 1966, 1978, 1981

| | EEO Firm Employees Distributed by Occupation | | | Normalized Shares | | | | | |
| | | | | White Male | | | White Female | | |
	1981	1978	1966	1981	1978	1966	1981	1978	1966
Managers & Admin.	11.7	10.8	8.2	1.54	1.56	1.47	0.55	0.48	0.32
Professionals	9.7	8.6	6.6	1.17	1.21	1.38	1.01	0.96	0.46
Technicians	5.7	5.0	4.5	1.10	1.10	1.09	0.98	0.98	0.99
Sales Workers	9.0	8.8	7.1	0.86	0.88	0.98	1.40	1.39	1.30
Clerical Workers	16.3	15.6	16.7	0.27	0.30	0.43	2.07	2.16	2.43
Craft Workers	12.1	12.6	14.2	1.63	1.58	1.46	0.21	0.21	0.20
Operatives	19.1	21.1	25.4	1.06	1.04	1.00	0.72	0.75	0.85
Laborers	7.5	8.5	9.7	0.92	0.89	0.86	0.70	0.74	0.66
Service Workers	9.1	9.0	7.7	0.61	0.60	0.66	1.15	1.21	1.13
	100.0	100.0	100.0						
% of EEO Employed				48.0	50.2	60.6	33.0	31.7	28.0

TABLE 4.2 continued

| | Normalized Shares | | | | | | | | | | | |
| | Black Male | | | Black Female | | | Hispanic Male | | | Hispanic Female | | |
	1981	1978	1966	1981	1978	1966	1981	1978	1966	1981	1978	1966
Managers & Admin.	0.47	0.41	0.12	0.27	0.21	0.08	0.50	0.48	0.31	0.27	0.21	0.10
Professionals	0.32	0.30	0.13	0.44	0.40	0.24	0.35	0.39	0.37	0.32	0.32	0.16
Technicians	0.63	0.57	0.26	0.89	0.92	1.03	0.68	0.68	0.53	0.59	0.63	0.56
Sales Workers	0.48	0.43	0.17	0.80	0.77	0.56	0.56	0.52	0.41	1.00	0.89	0.92
Clerical Workers	0.35	0.30	0.16	1.73	1.65	1.05	0.29	0.29	0.31	1.73	1.68	1.39
Craft Workers	1.20	1.10	0.55	0.22	0.21	0.17	1.38	1.32	1.01	0.27	0.32	0.33
Operatives	1.68	1.67	1.46	1.09	1.12	0.98	1.41	1.39	1.30	1.18	1.16	1.15
Laborers	2.08	2.06	3.07	1.16	1.21	1.45	2.47	2.48	2.79	1.73	1.84	1.74
Service Workers	1.65	1.60	2.32	2.20	2.35	3.93	1.47	1.35	1.62	1.59	1.58	1.57
% of EEO Employed	6.0	6.3	5.7	5.5	5.2	2.5	3.4	3.1	1.7	2.2	1.9	0.8

Note: The first three columns of the table show the distribution of all EEO firm employees (all sex and race combined) by occupations for 1966, 1978, and 1981 respectively. These three columns give an indication of the changing relative importance of the major occupations in EEO reporting firms in 1966, 1978, and 1981. The normalized shares of major groups of workers shown for 1966, 1978, and 1981 in the remainder of the table are computed by dividing the share of employment held by each group in each occupation by that group's share of all employment in EEO reporting firms (shown on the last line of the table). An index below 1.00 indicates underrepresentation; above 1.00, overrepresentation.

Source: U.S. Equal Employment Opportunity Commission, *EEO-1 Report on Minorities and Women in Private Industry,* 1966, 1978, and 1981.

a way to keep entry restricted. Still, the rising importance of education is also a reflection of a growing reliance on externalization of training. This tendency has been in the making for several decades, especially among the upper echelons of the occupational structure. Still, the new technology is intensifying the trend by accelerating the formalization of training and education for workers employed in a broad range of middle-level occupations.

The increasing importance of education appears to be creating both new opportunities and potential problems for groups of workers that have traditionally been the target of discrimination. On the one hand, externalization of training may make it increasingly difficult for employers to close off access to skill acquisition as a way to discriminate against women and minority workers. The great progress of women over the past two decades in professional occupations attests to this fact.

On the other hand, the externalization of training is unlikely to be free of problems. For example, the current structure of the higher educational system—characterized by considerable disjunction among various levels (two-year colleges, four-year colleges, graduate schools) and often lacking flexibility—makes it difficult to solve problems involving workable continuing education, which may be increasingly demanded in order to progress upward in the labor market during one's work life. In addition, to the extent that employers may partly control access to higher education—for example, by financing retraining programs at the community college or four-year college level or by financing tuition reimbursement programs—there may be room for discrimination to creep back in.

In general, these developments point to the increasing importance of the issue of who gets access to preferred education and why. In a society that is still far from having an equitable educational system in place, a family's socioeconomic status may largely determine one's future position in the labor market. It is no secret that over the past two decades it has been primarily middle-class women—mostly although not exclusively white—who have been most successful in advancing to higher occupational positions through the higher educational route. Short of major changes, this trend may accelerate.

In concluding, three points must be emphasized. First, I believe that EEO's emphasis on eliminating cultural biases and institutional arrangements that perpetuate discrimination in the workplace must be maintained. But I also believe that EEO must begin to reach outside the employing institution to the educational process in order, ultimately, to strengthen enforcement in the workplace.

Second, the linkage between work achievement, education, and socioeconomic background may have major implications for women's and minority groups that have traditionally used "sex " or "race" as a lever in the workplace. Increasingly, women and racial groups may become differentiated along socioeconomic class lines, so that recourse to "sex" or "race" as rallying points in the workplace may lose some strength.

Finally, the aggregate data call for special concern with regard to minority men, who appear to be failing to enter many of the white-collar occupations where much of the future lies. The unrelenting high unemployment among young minority male workers is a forerunner of serious social instability unless we find ways of intervening and succeed in turning the trend around.

Notes

This chapter is a slightly edited version of a paper by Thierry J. Noyelle, "The New Technology and the New Economy: Implications for Equal Employment Opportunity," prepared in 1985 for the Panel on Technology and Women's Employment of the National Academy of Science.

1. Thierry J. Noyelle, "Work in a World of High Technology: Employment Problems and Mobility Prospects for Disadvantaged Workers," paper prepared for the Educational Resources Information Center, U.S. Department of Education and the Office for Research in High Technology Education, University of Tennessee, Knoxville, 1984; Thomas M. Stanback, Jr., and Thierry J. Noyelle, *Cities in Transition* (Totowa, N.J.: Rowman & Allanheld, 1982); Thomas M. Stanback, Jr., Peter J. Bearse, Thierry J. Noyelle, and Robert A. Karasek, *Services/The New Economy* (Totowa, N.J.: Rowman & Allanheld, 1981); and Eli Ginzberg and George Vojta, "The Service Sector of the U.S. Economy," *Scientific American* 244:3 (March 1981).

2. U.S. Bureau of Labor Statistics, *Employment and Earnings* (Washington, D.C.: GPO, various years), household survey data.

3. Ibid., establishment survey data.

4. Ibid., household survey data.

5. Richard Edwards, *Contested Terrain* (New York: Basic Books, 1979); Paul Osterman, "Internal Labor Markets in White Collar Firms," Working Paper #90, Department of Economics, Boston University, September 1982; Paul Osterman, "Employment Structures Within Firms," *British Journal of Industrial Relations*, 1983; and Thierry J. Noyelle, *Beyond Industrial Dualism: Market and Job Segmentation in the New Economy* (Boulder, Co.: Westview Press, 1986).

6. Eileen Appelbaum, "The Impact of Technology on Skill Requirements and Occupational Structures in the Insurance Industry, 1960–1990," Report, Department of Economics, Temple University, April 1984; Barbara Baran, "The Technological Transformation of White Collar Work: A Case Study of the Insurance Industry" (Washington, D.C.: National Academy of Science, Panel on Technology and Women's Employment, 1983); and Noyelle, *Beyond Industrial Dualism.*

7. Appelbaum, "The Impact of Technology"; and Noyelle, *Beyond Industrial Dualism.*

8. Noyelle, *Beyond Industrial Dualism.*

9. Carmenza Gallo, "The Construction Industry in New York City: Immigrants and Black Entrepreneurs," Working Paper, Conservation of Human Resources, Columbia University, 1983.

10. Herbert R. Northrup and John A. Larson, "The Impact of the AT&T-EEO Consent Decree," Labor Relations and Public Policy Series, The Wharton School, University of Pennsylvania, volume 20, 1979.

11. C. Ichniowski, "Have Angels Done More? The Steel Industry Consent Decree," *Industrial and Labor Relations Review* 36:2 (January 1983).

12. Noyelle, *Beyond Industrial Dualism.*

13. Ibid.

14. Larry Hirschhorn, "Information Technology and the Office Worker: A Developmental View," Working Paper, Management and Behavioral Center, The Wharton School, University of Pennsylvania, 1984; Olivier Bertrand and Thierry J. Noyelle, "Development and Utilization of Human Resources in the Context of Technological Change and Industrial Restructuring: The Case of White Collar Workers," Expert Report, Organization for Economic Cooperation and Development, Center for Educational Research and Innovation, Paris, November 1984; and Appelbaum, "The Impact of Technology."

15. U.S. Bureau of Labor Statistics, *1995 Industry-Occupational Employment Outlook* (Washington, D.C.: GPO, 1984).

16. Paul Adler, "Rethinking the Skill Requirements of New Technologies," Working Paper HBS 84-27, Harvard Business School, Division of Research, Cambridge, 1984; Hirschhorn, "Information Technology;" A. Rajan, "New Technology and Employment in Insurance, Banking and Building Societies: Recent Experience and Future Impact" (Gower: Aldershot, Institute of Manpower Report Series); and others reviewed in Bertrand and Noyelle, "Development and Utilization of Human Resources."

17. Eileen Appelbaum, "Alternate Work Schedules and Women" (Washington, D.C.: National Academy of Science, Panel on Technology and Women's Employment, 1985); and Noyelle, *Beyond Industrial Dualism.*

18. Hirschhorn, "Information Technology."

19. Appelbaum, "The Impact of Technology."

20. Thomas Bailey and Roger Waldinger, "New York Labor Market: Is There A Skill Mismatch," *New York Affairs* (Fall 1984).

21. Noyelle, *Beyond Industrial Dualism,* especially Chapter 4.

22. Ibid.

5

Metropolitan Economies

Introduction

After a quarter century of rapid advances in computer and related technology, abundant evidence indicates that the pace of change, far from slowing down, is probably accelerating. This trend gives the promise of a broadening scope of applications in the private and public sectors and in the home. How is the role of metropolitan economies within the larger national economy being changed by these developments, and what new problems and opportunities are being created?

In answering these questions, three propositions are set forth: (1) The impact of the new technology cannot be understood without recognizing the nature of the forces that have brought about the rise of services—an economic transformation in which technology has played an important but only contributory role and that has, in itself, altered the role of metropolitan economies; (2) Although adoption of computer-related technology throughout the U.S. economy today is uneven, there is evidence that the nation has entered a new phase of rapidly accelerating and increasingly sophisticated applications that will involve a broad spectrum of users; (3) The impact of the new technology involves changes in the nature of work and the organization of user firms with important implications for the location of economic activity and, accordingly, for metropolitan economic development.

The Rise of Services[1]

During the postwar years services have grown from 57 percent of employment to more than 70 percent today, which accounts for virtually all job increases since the beginning of the 1970s.[2] This rapid transformation has not involved, however, all services equally. The expansion of service employment has been accounted for chiefly by the nonprofit services (education and health), government, and, although not widely appreciated, the producer services (finance, insurance, real estate, and other business services such as law, accounting, and advertising). Closely associated with the latter but not revealed directly by the employment data has been a rapid growth of producer service–like activities performed within corporate organizations of goods and service firms alike.

Similar rates of employment growth have not been experienced in the distributive services (wholesaling, transportation, communications, and utilities), although there has been no decline in their relative importance when measured in value-added terms. Retailing has grown more or less in step with the overall economy, and consumer services have declined somewhat, largely because of reductions in private household services. (There have been sharp increases elsewhere within the sector.) The goods-producing sector (agriculture, mining, manufacturing) has, of course, declined in terms of employment share; but it has not declined in value-added terms, which is a reflection of major advances in productivity.

Factors Contributing to the Rise of Services

The rise of services has involved a major transformation in both what is produced and how it is produced. In terms of final products (what is produced) there has been a marked trend toward greater product differentiation and proliferation of new products and services, reflecting higher levels of income and changing tastes. But the growth in government and nonprofit services has been the principal source of change in the final demand for services. In spite of some retrenchment in the public sector in recent years, it is clear that the affluent, complex modern society requires major flows of health, education, regulatory, and other services.

The transformation in how goods are produced has involved, on the one hand, changes of production in manufacturing and, on the other, a rising importance of producer services (both freestanding and in-house). Changes in manufacturing production have been made possible by improvements in factory technology, transportation, communication, and management science and have brought a reduction in blue-collar production employment and a continuous shifting of plants away from cities and (more or less in stages) into suburbs, smaller urban places, rural areas, and overseas. At the same time, there has been a shift in the importance of functions other than direct production, involving greater emphasis on finance, marketing, product development, personnel management, and problems related to coping with a more complex regulatory environment. It is this shift in functional emphasis that accounts for the rise in producer services and for the shift of resources toward headquarters, sales, and other service-like activities within the firm.

A number of forces—among which the rise of computer and related technologies is but one—have driven the changes sketched above. The increasing size of markets, made possible not only by a growing population and per capita income but by a breaking down of regional barriers and new trends toward internationalization, has opened up new demands for product differentiation, branding, and sales promotion and has brought about a host of new products. Improved transportation and communications have facilitated the delivery of products to these broad markets through more elaborate and more efficient distribution systems and have made it economic to locate manufacturing plants in nonmetropolitan areas.

The rise of the large corporation requires special comment as a major factor in the service transformation. There has been so much said in recent years regarding the importance of small businesses in generating job increases that the role of large corporations has tended to be downplayed. Large companies have not changed substantially in importance, in value-added terms, since the 1960s.[3] They hold center stage in any analysis of the impact of the rise of services and changing technology on metropolitan economies. It is the large corporation that has played the principal role in reaching out to national and international markets as well as in providing major shares of banking, insurance, transportation, and communications services. Moreover, in terms of the economies of metropolitan areas, large corporations play a strategic role. The siting of their activities influences the export base of the host city, both directly through adding to employment and income and indirectly through attracting support services (including, especially, producer services) and by enlarging demands for residentiary activities.

Finally, computer and telecommunications technologies have played an important role in the services transformation. Application of these technologies has made it possible for large corporations to take on an expanded role by enabling them to process vast quantities of data and to rebuild corporate structures in such a way as to organize and administer the new production, marketing, and developmental efforts. But technology has also made it possible for certain medium-sized and even small producer service firms (e.g., consulting, engineering, and advertising) to provide for the growing needs for expertise and specialized services among both private sector firms and governmental agencies.

The Changing Role of Metropolitan Economies

One of the major contributions of urban geographers and economists has been to make explicit the role of the export base in the economies of cities and towns. Urban places must export goods or services in order to import in turn, and labor, management, and other factors of production must be engaged in these activities as well as in local sector (residentiary) activities that service the local populace.

Cities have traditionally played a role as centers for providing distributive, governmental, and nonprofit services to their hinterlands, but the relative importance of these export roles has varied widely among places. During the era of industrialization, many cities both large and small became manufacturing centers, with major portions of their work forces devoted to production or related activities, while other cities continued to play roles as distribution centers for regional agricultural economies or as government centers.

As services have increased in importance and as goods production has been relocated, many cities that previously specialized in manufacturing have faced major problems of adjustment. All cities have faced major changes in their industrial structures as certain types of service activities have taken on increased importance.

In short, the growth of services has brought about a new competition among metropolitan economies, and some cities have clearly been favored over others. Evidence of such competitiveness can be gleaned from an earlier study that has been updated with recent data. In this study by Thierry J. Noyelle and myself, the 140 largest standard metropolitan statistical areas (SMSAs) were classified according to the industrial composition of employment and key characteristics of specialization in terms of business services, corporate headquarters, distribution, communications, and transportation.[4] The principal groups of SMSAs are as follows:

1. Nodal
   a. National nodal (New York, Los Angeles, Chicago, San Francisco)
   b. Regional nodal (e.g. Philadelphia, Houston, Atlanta)
   c. Subregional nodal (e.g. Memphis, Syracuse, Charlotte)
2. Functional nodal (e.g. Detroit, Hartford, Rochester)
3. Government-education (e.g. Washington, Albany, Harrisburg)
4. Education-manufacturing (e.g. New Haven, South Bend, Ann Arbor)
5. Production
   a. Manufacturing (e.g. Buffalo, Flint, Greenville)
   b. Mining-industrial (e.g. Tucson, Duluth, Charleston, W.Va.)
   c. Industrial-military (e.g. San Diego, Norfolk, Huntsville)
6. Residential-resort (e.g. Nassau-Suffolk, Tampa, Las Vegas)

Nodal places are service centers in which exports are concentrated primarily in distributive and producer services and often secondarily in other services as well (e.g. nonprofit services, arts, or recreation).

TABLE 5.1
Average Annual Rates of Employment Change: Total Employment and Selected Industry Groups, by Type of Metropolitan Economy, 134 Largest SMSAs[a] (1976–1983)

Type of SMSA[b]	Total	Mfg.	Whole-retail	FIRE	Other Services[c]	Government
National nodal (4)	1.37	−.92	1.45	3.36	4.13	−.16
Regional nodal (19)	2.64	.73	2.84	3.61	4.85	1.24
Subregional nodal (16)	2.20	.08	2.34	2.81	4.79	1.56
Functional nodal (24)	1.04	−1.50	1.85	2.62	4.12	.44
Government-educ. (14)	2.50	1.18	3.23	4.37	5.04	1.17
Educ.-mfg. (5)	1.36	−1.31	1.95	2.34	4.31	1.27
Manufacturing (22)	.63	−1.55	1.64	2.91	3.77	.38
Industrial-mil. (12)	3.14	2.04	3.91	4.12	5.90	1.42
Mining-industrial (12)	1.06	−1.25	2.32	3.50	4.30	1.20
Residential-resort (11)	4.95	4.38	4.90	6.70	6.63	2.42

[a] All 134 SMSAs are among the 140 SMSAs classified by Thierry J. Noyelle and Thomas M. Stanback, Jr., *The Economic Transformation of American Cities* (Totowa, N.J.: Rowman & Allanheld, 1984). Data for six SMSAs were not available. Annual rates shown are modified averages (highest and lowest values dropped in each classification, except for national nodal and educ.-mfg., where straight averages were computed).
[b] Numbers in parentheses indicate number of SMSAs.
[c] The principal components of the "other services" industry group are business services and nonprofit services (largely education and health).
Source: Compiled from U.S. Bureau of Labor Statistics, *Employment and Earnings,* June 1977 and June 1983.

Many, but by no means all, are state capitals and thus are also heavily engaged in delivering public sector services. Usually, the degree of diversification and specialization among nodal centers is a function of their population size and the size of the markets they serve. The presence of corporate headquarters or other administrative installations is of considerable importance in explaining the structure of their economy. The classification of these nodal centers under three headings—national, regional, and subregional—captures much of the variation that results from market size differences.

Functional nodal centers are places specialized in both manufacturing production and selected service functions of the large corporation, mostly R&D and administration of large industrial divisions. Although they can be quite large—many are comparable in size to the regional and subregional centers—they are much more restricted in their service functions than the nodal centers.

Government-education places are, for the most part, state capitals, seats of large educational institutions, or both. Education-manufacturing places are predominantly old industrial centers that are also sites of university complexes. Production centers include manufacturing, mining, and industrial-military places. They are characterized by more routine production than is found in functional nodal places and are singularly weak in attracting administrative or research establishments of industrial corporations. Industrial-military places differ significantly from other production centers in that they are characterized by a large presence of federal government, civilian, and military employees on military bases, shipyards, etc.

Finally, resort-residential centers include some of the metropoles that out-lie the large national nodal centers as well as many places that have developed since World War II as resort and retirement centers.

Comparison of the typical growth experience of each group sheds considerable light on what has been taking place. The rankings of the several groups of SMSAs according to average net employment growth are shown below for 1976–1983 (based on Table 5.1) and for the preceding period, 1969–1976:

	1969–76[a]	1976–83[b]
National nodal	9	6
Regional nodal	4	3
Subregional nodal	6	5
Functional nodal	7	9
Government-education	3	4
Education-manufacturing	5	7
Manufacturing	10	10
Industrial-military	2	2
Mining-industrial	8	8
Resort-residential	1	1

[a]Based on data analyzed in Thierry J. Noyelle and Thomas M. Stanback, Jr., *The Economic Transformation of American Cities* (Totowa, N.J.: Rowman & Allanheld, 1984).
[b]Based on Table 5.1.

An initial observation is that in terms of rates of change in total employment, the experience of the several groups has shown considerable variation. By far, the most rapid growth has been experienced by places that have been favored by the new postwar patterns of agglomeration: people-centered activities and the rise of new light industries (resort-residential), military build-up (industrial-military), activities related to corporate offices and business, financial and distributive services (regional and subregional centers), and activities strongly influenced by higher government or education-centered development (government-education places). The slowest growing places are those that are still most clearly identified with the earlier industrial era: manufacturing, functional nodal, and mining-industrial places.

But the aggregate employment data do not tell the whole story. The national nodal, functional nodal, and education-manufacturing cities have experienced heavy losses in manufacturing employment (Table 5.1) and posted aggregate gains only as a result of very substantial increases especially in business, nonprofit services, and FIRE services. Moreover, all of the national nodal and many of the functional nodal centers are so large that growth rates fail to reflect the extent of their vitality. Their modest net growth rates represent a considerable measure of success in transforming their economic structures. On the other hand, the manufacturing centers have not only shown little growth, but have, for the most part, been unsuccessful in bringing about a transformation toward a service-based economy. Growth rates of their small FIRE and other services sectors (Table 5.2) are the lowest of all the groups (Table 5.1).

Still further insights can be gained by observing the relative shares of employment accounted for by manufacturing and selected service categories and the changes in these shares that have occurred since the mid–1970s (Table 5.2). It is immediately apparent that the nodal centers and the residential-resort, education-manufacturing, and government-manufacturing centers have, relatively speaking, the most developed service sectors when measured in terms of combined shares of FIRE and others services (which include the financial, business services, and nonprofit services infrastructure) and that the production centers are least well endowed.

In short, the evidence suggests that comparative advantages and disadvantages are deeply rooted and that the weakest places are experiencing little success in effecting the kind of industrial transformation necessary to restore them to economic health.

Growth and Significance of the New Technology

The modern computer can be traced back to the ENIAC in 1946 but appears to have first found useful commercial application in the late 1950s. Since its inception there have been a series of breakthroughs in technology, which made possible first the large high-speed mainframe hardware and then, successively, the minicomputer and microprocessor, with ever more impressive advancements in computing speed, memory, and cost efficiency in each. There have also been other technologies

TABLE 5.2
Average Shares of Employment, Selected Industry Groups, by Type of Metropolitan Economy, 134 Largest SMSAs, 1976 and 1983[a]

Type of SMSA[b]	Manufacturing		Whole-Retail		FIRE		Other Services[c]		Government	
	1976	1983	1976	1983	1976	1983	1976	1983	1976	1983
National nodal (4)	20.9	17.9	21.0	22.0	8.4	9.6	21.1	25.4	17.0	15.2
Regional nodal (19)	20.7	18.3	24.1	24.6	6.6	7.1	19.6	22.7	16.9	15.3
Subregional nodal (16)	17.7	15.0	24.3	24.6	6.9	7.3	18.1	21.6	19.6	18.4
Functional nodal (24)	32.7	27.5	20.7	21.8	4.5	5.1	17.0	20.9	15.4	14.4
Government-educ. (14)	16.2	14.5	20.1	21.1	5.1	5.8	17.8	20.9	31.3	28.7
Educ.-mfg. (5)	26.7	22.1	20.4	21.1	4.6	5.0	19.2	23.2	21.4	21.3
Manufacturing (22)	37.1	31.9	20.3	21.8	3.8	4.3	15.9	19.5	14.4	14.3
Industrial-mil. (12)	17.6	16.4	21.4	22.7	4.2	4.5	16.9	20.2	28.9	25.7
Mining-industrial (7)	13.1	12.8	23.4	24.3	4.2	4.6	17.7	21.3	22.7	21.0
Residential-resort (11)	13.3	13.0	25.8	25.6	5.6	6.1	23.2	26.0	20.2	16.8

[a] All 134 SMSAs are among the 140 SMSAs classified by T.J. Noyelle and T.M. Stanback, Jr., *The Economic Transformation of American Cities* (Totowa, N.J.: Rowman & Allanheld, 1984). Data for six SMSAs were not available. Annual rates shown are modified averages (highest and lowest values dropped in each classification, except for national nodal and educ.-mfg., where straight averages were computed).

[b] Numbers in parentheses indicate number of SMSAs.

[c] The principal components of the "other services" industry group are business services and nonprofit services (largely education and health).

Source: Compiled from U.S. Bureau of Labor Statistics, *Employment and Earnings,* June 1977 and June 1983.

that have had parallel development, which are increasingly being joined to provide highly versatile and cost-efficient new configurations of equipment and applications. These other technologies include video, facsimile transmission, television, and, most important, telecommunications.

How does one begin to understand where these new, powerful technologies are leading and to assess their likely impact on the U.S. economy in general and the U.S. city, in particular? In my view, such understanding requires that one recognizes (1) that adoption of computers and related technology has proceeded unevenly; but (2) that virtually every branch of the economy is being affected by a new era of applications of computer and related technology; (3) that the new technology is bringing about major changes in the way work is done and the way firms are organized; (4) that these changes are creating new locational options for firms and nonprofit organizations; and (5) that the special requirements of telecommunications networks in this new era of applications are likely to influence the competitiveness of firms and urban places and to create new opportunities for some and obstacles for others.

Unevenness of Application

As the technology has evolved, the extent to which it has been applied has varied widely among users. Such unevenness is due to a number of factors: The nature of the firm's product (or mission), the size of the user organization, and factors surrounding the investment decisions, such as financial constraints and the progressiveness of management, will all influence the extent to which a particular firm will apply the new technology.

In some organizations the new technology is essential to the firm or nonprofit organization because it makes possible certain strategically important service features or brings about radical reductions in the cost of production. Where either obtains, organizations tend to move quickly to adopt the technology. When AT&T first offered private line data communications facilities in 1958 the major airlines promptly installed reservations systems, and stock market quotation facilities began to be put into place only a few years later. Similarly, computer applications have permitted banks to move rapidly toward new forms of competition.

The size of the user organization also plays a role, of course, with large organizations in any given industry tending to be the earliest users as well as the most sophisticated. Not only is the technology most effective when applied to the processing of large quantities of data, the handling of heavy telecommunications-oriented traffic, and the monitoring and coordinating of dispersed activities, but large organizations are also more likely to have the financial means to invest in necessary equipment and software and to dedicate the staff to the planning and supervision involved.

Although large organizations have led the way, small organizations have not been barred from making use of the computer. From the outset service bureaus made available the capabilities of large mainframes and the expertise of trained staffs for a variety of fairly routine computational needs such as billing, payroll, and inventory accounting, largely on a batch process basis. However, small firms have typically lacked the finances or staff to purchase hardware and software and to create and use more complex data processing systems. In recent years the availability of inexpensive microcomputers along with off-the-shelf software, particularly spread-sheet capability, has opened up opportunities for bringing the computer into the small organization, not only for basic records processing but also for managerial control and planning.

A New Era of Computerization

What appears to be taking place is a rapid transition from an era in which computers were used principally as adjuncts to work done in the various administrative departments to an era in which they are used for a vast array of managerial tasks including organizational control, planning, and financial and marketing strategies. In the new era the necessary technologies are sufficiently advanced in sophistication and cost-effectiveness and software, along with professional expertise, is sufficiently available that users are finding themselves under great pressure to adopt the computer or to expand its use. Moreover, the accumulated experience with earlier applications coupled with a new generation of management has ushered in a new environment in which change is likely to be favored rather than resisted. Above all, it is an era in which the several related technologies, especially computers and telecommunications, have been combined to open up entirely new uses, which involve, increasingly, the inauguration of integrated systems and the building of data bases accessible throughout the firm without the earlier restrictions upon space or function.

Richard L. Nolan, a leading computer management consultant, has provided a useful model that sheds light on what is currently taking place by describing how computer processing develops from initial adoption to most advanced application in a prototype large corporation.[5] The six stages that he delineates fall into two "eras." The "computer era" includes the stages of (1) initiation, (2) contagion, and (3) control, and the "data resource era" includes the stages of (4) integration, (5) data administration, and (6) maturity. Progress from stage to stage is driven by the dual effect of the accumulation of experience within the user organization and rapid improvements in technology. During stages one and two "there is a concentration on labor intensive automation, scientific support and clerical replacement," during stages three and four "applications move out to user locations for data generation and data use," and during the final stages "balance is established between centralized shared data/common systems applications and decentralized user-controlled applications." The computer era begins with 100 percent batch processing and moves toward introduction of data base processing and inquiry and time sharing processing. In the data resource era, batch and remote job entry processing are progressively reduced, data base and data communications processing becomes the dominant mode (well over 50 percent of computer usage), and personal computing and minicomputer and microcomputer processing together account for over a fourth.

Nolan sums up where the typical large corporation stood at the beginning of the 1980s as follows:

> . . . We've spent the last 20 years learning how to formally define systems, and install systems analysis programming . . . to provide data processing support to business functions. We've done that with a technology which we generally call the batch technology and, in a real sense, we have faced severe limitations in using that (technology) robustly. . . .

> . . . Now what happens in the Stage Three-Four environment is that organizations retrofit those applications . . . to permit them to tap the data resource much more easily. Secondly, organizations bring capability forth to the user (within the organization) through terminal technology, minicomputers and distributed systems. When these two things are fitted together, the results are that 20 years of investment are suddenly unleashed, results that have been pretty well locked up in until-now inflexible systems. These dynamics seem to be what's driving the Stage Four growth that we see. So much latent user demand has been pent up in inflexible systems in earlier years that freeing it causes a near explosion.[6]

There seems to be solid evidence that we are indeed entering a new era of computer usage. National Income Division estimates indicate that investment in computer and telecommunications technology—defined as office, computer and accounting machinery, and communications equipment—rose from 18 percent of total nonresidential durable equipment expenditures in 1976 to 37 percent in 1982. Still further evidence is found in Nolan's data, which show that data base management software installed on IBM medium- to large-scale computers in the United States rose from about 15 percent of such equipment in 1975 to roughly 80 percent in 1980.[7]

In interviews with users, my colleague Thierry J. Noyelle and I have found large banks and insurance companies and several of the large department store chains to be far advanced in the conversion toward integrated systems. New York City is moving rapidly toward development of an elaborate telecommunications network and advanced systems of monitoring and administering a variety of the City's services (though by no means all). We found relative backwardness in the computerization of administrative procedures of two major universities, although in each plans are under way for major advances in the years just ahead.[8] Even casual reading of both the technical journals and the business press provides additional examples of acceleration and rapidly broadening application of the more sophisticated technology available today.

But the acceleration of the application of computers to sophisticated uses is not limited to the installation of distributed processing systems. Among both large and small users the availability of microcomputers at extremely low prices, along with a variety of easily used software, is setting off a virtual revolution in terms of new applications. In large organizations departments are refusing to wait for the more complex systems and are putting into place subsystems using desk computers to monitor, analyze, and plan their own operations. We have encountered this not only among large firms but also within universities and the various agencies of the City of New York.

In small business there has been a rush to apply the new hardware to a variety of uses such as inventory control, routing and monitoring salesmen, planning and controlling cash flow and investment, filing records, and, of course, word processing.

The applicability of the microprocessor is astonishingly broad and the impact in terms of efficiency and sophistication of operation is very impressive indeed. This new development raises questions regarding the role of small businesses (both services and manufacturing) and professional organizations. It is entirely possible that large firms, governmental agencies, and nonprofit firms will increasingly find it feasible to contract out additional functions to these smaller organizations and that smaller businesses, given access to telecommunications networks, will be able to expand the geographical scope of their markets.

Changing Work and Organization

The evolution of computerization has involved both a change in the way work is done and a change in functional emphasis within the user firm. On one hand, there has been a change in the work content of occupations and in the relative importance of occupations; on the other hand, there has been a change in emphasis given to various functions and to the allocation of resources within the user organization.

From the outset, computerization made possible the reduction or elimination of some tasks (e.g. filing, laborious computations, and preparation of forms) and created others (e.g. programming, computer operations and maintenance). At the same time it freed up budgets so that the emphasis on marketing, product development, and management could be increased. As applications have become more sophisticated and particularly as data base systems have come into use, the computer has increasingly been used by management to provide a more carefully customized line of products or services, to control finances, to redesign sales and promotional efforts, and to improve planning and managerial control—all of which are bringing about changes in work and in the organization of the firm.

Organizational changes have occurred at every level. Distributed processing has permitted much of the routine processing to be shifted out of general computing centers and back to the originating departments. Where high-volume processing tasks such as check clearance, accounts receivable, or credit card processing have continued to require large cadres of clerical personnel, the availability of telecommunications systems has made it possible to relocate the clerical staff away from the headquarters operation to sites where rents are lower, labor supply is more suitable, and wage rates are more favorable.

In the industries that we have studied there have been marked changes. For the large department store with multi-branch operations, the installation of point-of-sale terminals has made it possible not only to transmit the information necessary for the consolidated processing of accounts receivable and billing to higher echelons, but also to have more control over inventory and purchasing. Along with the latter, the buying functions have shifted to the central office. With headquarters assuming more of the merchandising as well as financial, general policy, and planning functions, the responsibilities of individual store management are increasingly being limited to selling and general housekeeping.

The large banks and insurance companies have been among the leaders in using sophisticated data processing, which has had sharp impacts on work and organization at all levels. Not only did computerized check handling reduce back-office staffs at an early stage, but more recently automated teller machines have reduced the use of regular tellers and permitted those who remain to handle a greater variety of transactions. Integrated systems with on-line enquiry have permitted account officers to retrieve information on their customers and to provide a greater range of services than before. Among insurance companies we have observed equally dramatic effects. Billing and premium accounting are becoming automated and centralized, and on-line enquiry into data bases (containing information formerly stored in massive filing systems) are making possible more efficient and prompt customer service. Problems of rate setting and submission of reports are much more readily solved than formerly as are problems that involve management of investment portfolios, and ready availability of information from central data bases is changing the role of salesmen and the functions of sales offices.

Thus, the general effect has been a redefinition of many jobs, a deletion of some, and an addition of others, with some departments taking on new importance and others less. An active controversy is abroad as to whether work is being upskilled or deskilled, but the terms of the debate appear to be misstated. Although it is clear that computerization has in the past been most successful in eliminating repetitious, low productivity work, it has also acted to simplify work of all sorts. What seems to be taking place is that, in a rapidly changing environment where even the lower levels of work often involve the use of expensive equipment and higher levels involve coping with a much wider range of problems and procedures than formerly (albeit with the powerful assistance of technology), employers are placing a greater emphasis than before on literacy and the ability to learn new ways of working. For high-level work there is also a new emphasis on credentialing through advanced business or technical training. At the same time the new world of work places penalties on those with poor educational backgrounds and levels of competence. Those low-level jobs that are open to them tend to be scheduled on a part-time basis, pay poorly, and offer few fringe benefits or opportunities for advancement.

There are also new trends in work arrangements as employers make use of part-time and flex-time work schedules to accommodate both the requirements of the new work and the needs of a changing labor force. For the most part the shift to white-collar work has released employers from the need to follow the old eight-hour staffing patterns of the factory. Through increased use of part-time scheduling to match the flow of work, payroll costs have been reduced in sales and to some extent in clerical occupations. On the labor supply side, there have been major changes in the composition of the labor force. Women have come to play an increasingly important role in the labor force, their share of jobs rising from 29.6 percent in 1950 to 43 percent in 1980. Working women are much more likely than they were in the past to be married (59.3 percent in 1980 compared to 52.1 percent in 1950) and to have children (62.5 percent of women labor force participants in 1980, up from 28.3 percent in 1950).[9] This growing importance of women in the work force has brought about new demands for flexible work schedules, more convenient transportation, and better security both on the job and while journeying to work.

Technology and Changing Locational Options

A major implication of these changes in work, organization, and the composition of the labor force is that firms are faced with new options and new constraints in locating their activities. For the large firm there are opportunities to strengthen some functions at general and divisional headquarters, to thrust some functions outward to sites where labor is cheaper and of a more suitable quality and where rents and taxes are more favorable than in the major cities, and to eliminate some operations altogether or to consolidate reduced operations in new sites. For small organizations, new alternatives may open up if data base technology and telecommunications make it possible to service broader markets. On the other hand, cost and supply factors that were acceptable at a given location under previous conditions may erode the firm's competitive position when new procedures and organizational arrangements are put into place.

These new options are not necessarily radical departures from the kinds of alternatives that have been available to many users for a number of years, but they are sufficiently different to bring about important changes over time in the location of service activities. Thus, shifts of headquarters and, particularly, back-office activities from city to suburb have been taking place for many years, but the new computer-telecommunications technology makes such relocation feasible for a larger spectrum of activities.

An educated labor supply appears to be emerging as a locational factor of major proportions. The occupational statistics show clearly that there is a rising demand for professional, technical, and managerial staff in both absolute terms and as a relative share of the labor force. Firms and nonprofit activities must draw upon a pool of such personnel and be able to offer the amenities to hold them. Clerical jobs also continue to increase with the rise of services, but not as a relative share of employment in many industries. The important observation regarding labor requirements for these workers—largely women—is that employers appear to be raising their hiring standards. As a result, the competitive position of suburbs (where there is typically a pool of educated housewives) relative to the central city may well be changing.

For many corporate headquarters, research and development facilities, engineering offices, and related service facilities, proximity to one or more major universities appears to be a decisive factor. This is seen readily in the rapid development of the centers of high-tech complexes, where light manufacturing, R&D, and a variety of engineering and service activities exist cheek by jowl. In the Silicon Valley (near Palo Alto), Route 128 (near Boston), The Research Triangle (North Carolina), and the new high-tech complex near Philadelphia, the proximity to major universities is clearly of fundamental locational importance.

But access to a well-developed infrastructure of business and financial and other key supporting services, especially major airport facilities, continues to be a basic requirement for many firms. Such services are the principal source of economic strength for the large central city because the mutual attraction of a variety of headquarters, distribution services, advanced business services, and educational and nonprofit institutions acts to hold existing firms and attract others. The trend toward locating headquarters and related establishments far enough away from the city to take advantage of lower rents, favorable labor supply, and less congested surroundings while remaining close enough to have access to key financial, business, and airline services affects the city's economy in turn, but just how much and how consistently from city to city is a matter about which very little is yet known.

The Role of Telecommunications

Finally, one must recognize the strategic role of telecommunications. Just as the far-flung network of railroads was the infrastructure that permitted the coming of the age of industrialization and the relocation of manufacturing and agriculture, so it is that the telecommunications system stands as the basic infrastructure of the new era of computer usage that makes possible the new locational patterns of economic activities that are developing.[10]

The most basic telecommunications system for data processing is the existing telephone network, which has been used from the outset. Its principal limitations are that the conducting medium (single copper wire) is limited in capacity and that the U.S. telephone system is only partially equipped with the expensive electronic digital switching equipment necessary for data transmission. Moreover, efficient high-volume data transmission requires not only the availability of such electronic switching but also the linking of computer hardware and terminals through coaxial cable, microwave radio (either satellite or terrestrial), or fiber optic cable.

The requirement of a highly capital intensive infrastructure creates significant inequalities of opportunity among users and among places in terms of their ability to make use of integrated systems over broad geographical areas. A number of very large corporations (e.g. Citibank,

Sears, and J.C. Penney) have invested enormous sums to put into place nationwide and even international systems linking their various establishments. But smaller users and users with less extensive requirements must be able to tie into established networks in order to make use of distributed data processing. For them the facilities of the local telephone system may be the limiting factor because lack of electronic switching, which is available for well under 50 percent of the national network, may foreclose extended use of sophisticated data processing systems, at least for some years to come.[11]

Sophisticated telecommunications systems, like modern jet ports, may be expected to develop most rapidly where traffic is heaviest. This development is likely to reinforce the initial comparative advantages of those places that are already the most advanced service centers or that are developing most rapidly and are likely to attract the earliest capital outlays. Thus, New York City, which is the largest and most important service center of the nation, is equipped with the most sophisticated telecommunications system. Its telephone system has been substantially modernized and for the entire business district, at least, converted to electronic switching. Manhattan is ringed by an optical fiber system and fairly crisscrossed with coaxial systems. Several satellite stations are located either within the City or are available to users through coaxial or fiber optic cable.[12] It is difficult to imagine that such facilities will not provide a major source of strength to the City's economy in enabling it to hold onto its position of preeminence in finance and commerce.

But telecommunications infrastructure requirements do not appear to offer the same impediments to dispersal of activities to the suburbs of large cities as to distant smaller cities or towns. Telephone networks are likely to be upgraded in large metropolitan areas before they are upgraded in smaller places at more distant points. Alternate systems that use advanced technology to link the suburbs with the central business district are becoming available throughout the large metropolitan areas.

Technology and Competition Within the American System of Metropolitan Places

A principal conclusion of the preceding analysis is that the accelerating application of computer and related technologies along with the rise of services is bringing about new possibilities for the location of firms and leading to competition among places for growth and development at two levels: among metropolitan areas and between cities and suburbs within given metropolitan areas. In each arena of competition the outcome will be influenced by locational attractions of labor force quality, availability of telecommunications networks, the infrastructure of business services, educational, government, and nonprofit institutions, and a number of quality of life factors.

The argument is sometimes put forward that the slow growth places, and particularly old industrial centers under stress, can be rescued by bringing in new service activities by means of the computer and telecommunications technology. This argument must be carefully reevaluated in light of this discussion. For these places the prerequisites—an appropriate labor supply, a minimum complex of supportive services, and adequate telecommunications—are likely to be lacking.

It may well be possible to revitalize these places, but radical therapy will probably be called for. Their comparative advantages lie for the most part in the area of production. Large infusions of capital to bring in new manufacturing technology could, at least in some places, reestablish a viable export base around which supporting service activities could flourish.

But there is a more general lesson to be learned regarding the implications of the new technology for urban growth and development. This lesson, to be drawn from all that has been said above, is that the new applications of the computer and its increasing adoption in all types of economic activity, private and not-for-profit sectors alike, place new demands on the labor force and the urban infrastructure. In spite of what the data suggest about the comparative advantages of certain types of places, these new demands call for new initiatives in all places that seek to maximize economic well-being. What is called for is an upgrading of educational and training institutions (including new arrangements for retraining workers at all stages of their careers) and improvements in counseling and employment services, transportation services, and public sector services relating to safety and health and to the care of the children of working mothers. A concerted effort must be made to improve the quality of life in such a way as to hold on to and attract establishments that are in many instances being faced with a much broader set of locational options than heretofore.

This general conclusion is particularly relevant for cities and suburban communities that must face the issue of developmental strategies under conditions in which the new technology is raising new choices of location for firms within the metropolitan area. It is not yet clear whether the new locational trends will on balance favor city or suburb. What does seem clear is that the matter is not altogether in the hands of the Gods. Appropriate and aggressive public policy can do much to strengthen the local economy.

Notes

This chapter is a slightly edited version of "Technology, Services and the Changing Future of Metropolitan Economies" by Thomas M. Stanback, Jr., a chapter in *High Technology, Space and Society*, edited by Manuel Castells (Beverly Hills, Ca.: Sage Publications, 1985). Reprinted with the permission of the publisher.

1. For a more extended analysis of the rise of services, see Thomas M. Stanback, Jr., Peter J. Bearse, Thierry J. Noyelle, and Robert A. Karasek, *Services/The New Economy* (Totowa, N.J.: Rowman & Allanheld, 1981).

2. The rising importance of services has by no means been restricted to the postwar years. During the period dating from roughly the last quarter of the 19th century until the Great Depression, the rise of the industrial economy was accompanied, and in many ways facilitated, by a rapid rise in distributive and certain producers' services such as banking and insurance. The postwar period has brought a different pattern of changes, however, and is our only concern here.

3. Candee S. Harris, *Small Business and Job Generation: A Changing Economy or Differing Methodologies?* (Washington, D.C.: U.S. Small Business Administration, February 25, 1983), Revised, pp. 4–5 and Table D.

4. Thierry J. Noyelle and Thomas M. Stanback, Jr., *The Economic Transformation of American Cities* (Totowa, N.J.: Rowman & Allanheld, 1984).

5. Richard L. Nolan, "Managing the Crisis in Data Processing," *Harvard Business Review*, March-April 1979, pp. 115–126.

6. "ICP Interview: Richard L. Nolan," *ICP INTERFACE Data Processing Management*, Summer 1980, p. 19.

7. Richard L. Nolan, "Managing the Crisis in Data Processing," p. 122.

8. For an enlightening and entertaining summary of the New York City experience, see Desmond Smith, "Brave New City Government," *New York Magazine*, May 14, 1984, pp. 56–64.

9. The shares of women with children are not strictly comparable because of data reporting changes but provide a rough approximation of the changes involved. See Anna B. Dutka, *New Types of Work Scheduling: The United States Experience*, Conservation of Human Resources, Columbia University, 1984, unpublished.

10. Gershuny and Miles have called attention to the importance of the telecommunications infrastructure in their recent analysis of the new services economy in the European economy, J.I. Gershuny and Ian Miles, *The New Service Economy: The Transformation of Employment in Industrial Societies* (New York: Praeger, 1983).

11. Distributed data processing systems can, of course, be established by large organizations for local areas using local telephone networks and privately owned switching equipment.

12. For a more detailed description of New York's telecommunication system see Mitchell L. Moss, "New Telecommunication Technologies and Regional Development," Report to the Port Authority of New York and New Jersey, 1984, unpublished.

6

New York City: Dangers Ahead

Introduction

These remarks draw on ongoing research into the U.S. economy's shift to the services, the impact of this shift on the nation's cities and labor markets, and the linkages between the new computer-communications technology and these structural shifts. There are five economic development issues that loom increasingly larger on the agenda of New York City: technology, international trade, financial services, skills and education, and social bifurcation. In each of these five areas there are signs of difficulties ahead that have not yet received the attention they deserve, largely because of the current euphoria about the City's remarkable recovery from the financial and economic crisis of the early and mid–1970s.

The Recent Past

When the next generation of historians looks back at the period of the 1970s and 1980s, it will probably agree that these two decades represented a major turning point in U.S. economic, social, and technological history and that New York City was often at the leading edge of this social transformation.

What we are witnessing today is the demise of the economic system of industrial mass production, which constituted the backbone of the U.S. economy from the 1920s until the early 1970s and was linked to the emergence of mass consumer markets in the post–World War II era. Today a new paradigm of economic development is emerging, one in which a society once permeated by a culture of blue-collar/factory work is defining new models around white-collar and office work; one in which highly standardized outputs of an earlier period are being replaced by increasingly customized goods and services made possible by the highly flexible new technology; and one in which the continuing functionality of the large, heavily centralized corporation of yesteryear is being questioned.[1]

A principal manifestation of this transformation can be seen in the economy's shift to services. As my colleagues and I at Columbia University have sought to explain,[2] the shift to a service economy does not mean that services are replacing manufactured goods any more than the emergence of the industrial society eliminated the production of agricultural commodities. Rather, it means that an increasing share of the value added contained in final goods and services is created less and less through direct production work, which is being increasingly automated, and more and more through indirect, mostly white-collar work centered in the research and development of new products, the design and monitoring of automated systems needed to produce goods and services, and the marketing and selling of products.

New York City's claim to success in the past fifteen years rests on its having gone the farthest of all the nation's cities in spearheading this transformation by shedding its old, mostly manufacturing-dominated export base and expanding its base of producer services that are at the core of the new economy.

Having succeeded in accomplishing this transformation, there is a danger that the City may have become somewhat complacent about its future.

In general, producer services fall into three broad groups:[3]

1. Strategic planning and financial services that are needed to plan and implement the development path of firms, nonprofit institutions, public sector agencies, and other key actors in the economy and that include many of the services provided by commercial banks, investment banks, insurance companies, accounting firms, law firms, and others;

2. Distributive services that are needed to link the outputs of the firm or institution with its market and that include the gamut of services from transportation and wholesaling to marketing and advertising;

3. Developmental and innovational services that are needed to develop new processes and outputs as well as improve old ones and that include services ranging from those provided by private or university-based R&D labs all the way to software firms or in-house systems divisions of firms that design new automated systems.

There were already signs in the 1970s that New York could not take for granted its hold on key services and that other cities were challenging it to gain larger shares of those services. Witness, for example, the steady exodus of large headquarters from the City, the relocation of many business services, and the decline of transportation and wholesales linked to the decline of New York City's harbor. More recently, the out-migration of the back offices of large corporations and large producer service firms has caused concern, but it has been largely mitigated by the general buoyance of the City's economy.

Five Trouble Points

The New Technology

There is a tendency today to believe that the worst of the new technology's displacement effects are over, but nothing could be further from the truth. Unbeknownst to most observers is the fact that while back offices were being moved out, some firms were also busy shifting their systems development and programming work to locations outside the City.

In today's economy, software development is becoming perhaps the most critical area of the entire product development strategy of firms and businesses. Software is currently developed by various participants, including not only commercial software firms, obviously, but also publishing houses, management consulting firms, accounting firms, telecommunications firms, computer manufacturers, and also large user organizations themselves—including, say, Citicorp, General Motors, and the U.S. Department of Defense. Estimates of the dollar volume, now a few years old, revealed that although advanced systems were increasingly being developed by highly specialized firms, the bulk of the dollars were still spent on in-house development by user organizations. In terms of total local spending, New York was far ahead of any other U.S. city.

My recent research in the financial industry, however, indicates that in such a major New York City industry as insurance the exodus of systems divisions to locations outside the City has already begun, typically to second-tier U.S. cities with strong educational institutions, that is, to cities that can provide both lower operating costs and a good and steady supply of skilled labor.[4] Furthermore, this trend has taken an even more dramatic turn over the past couple of years, having reached a point where some New York City financial firms are contracting out computer programming and low-level systems development work to software firms located in developing countries.

These illustrations indicate that New York City is not only facing competition for its clerical functions, but is increasingly facing competition for high-level, skilled professional work. If the logic of the new technology is to automate a considerable amount of low-skilled work, then it is clear that tomorrow's competition among cities and across nations will be centered increasingly on skilled work. The model of the footloose assembly plant in the semi-conductor or apparel industries moving about from New York or California to Hong Kong or Thailand is not the appropriate model for competition in tomorrow's service economy. Rather, it is the model of R&D laboratories being shifted to Israel or of U.S. software work being farmed out to firms in India. Such a scenario has ominous implications for New York City.

International Trade

Another potential competitive threat to New York City's economy lies in the fact that the center of gravity of the world's economy is clearly shifting from the Atlantic to the Pacific. In 1982, for the first time in U.S. history the share of Pacific trade was larger than that of Atlantic trade. The gap is widening and will pose new challenges to the City's business institutions, which have tended, in the past, to have more of a Western European or Middle Eastern orientation than an Asian orientation.

Consider the Far East. The more than 700 million Indians, the 1.2 billion Chinese, the 120 million Japanese, and the several hundred million inhabitants of Indonesia, Malaysia, South Korea, Pakistan, and other countries of the region constitute formidable potential markets. But most of these countries have their eyes turned primarily toward Japan, Hong Kong, and Singapore, much less toward the United States and Western Europe. The economic development in that part of the world, which is steadily accelerating, is certain to leave its mark on the West, and it is far from clear how New York City will respond to the new challenge.

Financial Services

A third concern lies in a potential weakness in the City's financial sector—a weakness that is the obverse of its great strength. New York City's financial sector is now overwhelmingly dominated by very large institutions whose strength lies principally in extensive international activities heavily geared to serving national governments and large corporate customers. Recently, with Citicorp, Merrill Lynch, and American Express in the lead, New York City-based institutions have also tried to use their newly acquired technological edge to move back aggressively into the consumer financial markets. But the City's financial institutions have long been weak in serving effectively the middle and small corporate market.

Yet, much of the recent economic growth seems to have been stimulated by the success of new small and middle-sized firms that have taken the lead in bringing new technologies and new products to the market. A traditional answer from local financiers is that New York City has provided nearly 40 percent of the venture capital market. Yet, as former

California Bank Supervisor Derek Pete Hansen indicated recently, this response overlooks the fact that the few billions of dollars of equity offered by venture capitalists pale in comparison to the more than $150 billion mostly working capital loans extended by commercial banks to small and medium-sized businesses:

> Our financial system does some things very well even on the investment side. It provides low cost, abundant capital to risk-adverse, mature, well-capitalized and collaterized businesses. These businesses meet the lending requirements of risk-adverse, highly leveraged and government-guaranteed financial institutions. At the other end of the scale, we have a newly vibrant venture-capital community that can finance those few high-tech, high-growth businesses that provide the necessary fast returns to support unleveraged funds. What we do not have is adequate financing for the many growing businesses that fall between the two; businesses too small, new, fast-growing, or aggressive to justify adequate bank credit, but not sufficiently upscale to attract venture capital. In most states, these are the businesses that have the greatest potential for growth and job creation.[5]

If New York City's financial sector is weak in its capacity to finance small and medium-sized businesses, there are two important implications. First, New York City may not always provide the optimum environment to benefit from new entrepreneurship. Second, if, as I believe, the United States must strengthen its international trading position by promoting trade by middle-sized firms, some of New York City's financial institutions may lack the required expertise both to promote such a trade and to benefit from it. They could encounter increasing competition from U.S. regional banks or even from European and Japanese financial and public sector institutions, which appear to be more advanced in addressing the special problems of trade financing and promotion for middle-sized firms.

Skills and Education

Our research on the changes taking place as a result of both the shift to services and the introduction of new technology point to upskilling in many areas of the economy. This upskilling is leading to higher standards in hiring requirements, reflected particularly in the growing preference of employers to hire new employees directly from higher education institutions.

Several implications follow. First, New York City will have to compete with other cities increasingly on the basis of the strengths and weaknesses of its local school and higher educational system. There are many factors that help to explain the recent loss of back-office employment in the City, but one of them, undoubtedly, had to do with the fact that some employers were becoming worried that the City's educational institutions were not able to turn out an adequate supply of skilled clerical and low-level professional labor needed to staff their offices.

A second implication involves the barriers that distort the flow of students into and out of the higher education institutions. Even in the absence of overt discrimination, the evidence is clear that the proportion of minorities who graduate from high school and complete college is far less than the proportion of whites. The few special remedial programs in place, although helpful, have not succeeded in attenuating the trend to any appreciable extent. This issue requires close attention.

The third implication is that those who are left behind by the educational system are likely to be in real trouble. The increasing disinclination of large employers to hire students who have not earned a high school diploma and do not possess reasonable competences in language, mathematics, and analysis bodes ill for many inner-city youth. Too many drop out of high school with less than minimal prerequisites for regular employment. The gap between the meager credentials these young people offer and what employers want is often too wide to be closed by short-term remedial programs. Clearly, the youngsters who fail are not to be blamed for their failure, the roots of which lie deeply imbedded in weak family structures, poverty, discrimination, and in-effective public schools. Daniel Chall's article in *Quarterly Review of the Federal Reserve Board of New York,*[6] highlights that New York City's problem with illiterates and dropouts is far worse than in most cities.

Although there are still jobs available that require little skill, there will probably not be enough of them because of the strong tendency toward upskilling. In the current context of transformation, we are not about to see a repetition of the scenario of the late 1960s when the economy ran full speed and unemployment rates dropped to a minimum. In a manufacturing economy, almost anyone was employable if she or he were willing to make the effort to work. In a service economy, this is no longer true.

Although some partnership programs have been developed in the City between educational institutions—primarily high schools or com-munity colleges—and private sector employers, these programs speak only to a small proportion of the student population that needs to be reached. These experiments say little about the massive effort that the City needs to engage in to assure that all young people will be able to get a job in the service economy.

Social Bifurcation

The impact of the current economic transformation on the social fabric of the City is also cause for concern. Admittedly, the concern is not unique to New York City, but it does have a much broader range there. The recent economic recovery has overwhelmingly benefited a new class of so- called "young urban professionals," but it has left many behind. This trend is leading to a new and serious social bifurcation. There are many factors at work, including displacement of workers from older industries, rising unemployment rates among blue-collar workers, lagging wages in older industries, rising youth unemployment, tendencies toward underemployment in some of the service industries, and tendencies for lagging wages among many low-skilled or even skilled service industry occupations. The simple fact, however, is that many people are finding it more and more difficult to make a reasonable living in the City.

It is not fashionable at present to give heed to those who believe that an economy can get into serious trouble through the lack of buying power of a large sector of the population. But when this imbalance occurred in the late 1920s, the Great Depression followed. We need to be concerned, in the mid–1980s, that we do not repeat the same mistake.

Notes

This chapter is a slightly edited version of "New York's Economy in the Year 2000: Storm Clouds on the Horizon?" a testimony presented by Thierry J. Noyelle in 1985 before the City of New York Commission on the Year 2000.

1. See for example, Eli Ginzberg and George Vojta, *Beyond Human Scale: The Large Corporation at Risk* (New York: Basic Books, 1985).

2. See for example, Thomas M. Stanback, Jr., Peter J. Bearse, Thierry J. Noyelle, Robert A. Karasek, *Services/The New Economy* (Totowa, N.J.: Rowman & Allanheld, 1981).

3. See author's paper, Thierry J. Noyelle, "The Shift to the Services, Technological Change and the Restructuring of the U.S. System of Cities," UNIDO, Regional and Country Branch, Conference Room Paper #15, Vienna, August 20–24, 1984.

4. See Thierry J. Noyelle, *The New Technology and Financial Services: A Paradigm for the New Economy* (Boulder, Colorado: Westview Press, forthcoming).

5. Derek Pete Hansen, Testimony before the U.S. Congress, House Committee on Banking, Finance and Urban Affairs, May 2, 1984.

6. Daniel Chall, "New York City's Skills Mismatch," in *Quarterly Review of the Federal Reserve Board of New York*, Spring 1985, Vol. 10, no. 1.

7

Directions for Policy and Future Research

The New Technology: Major Employment Impacts

We became interested in the impact of the computer-communications technology on the U.S. labor market initially because we believed that the new technology would bring about significant changes in the job and career opportunities for women and the disadvantaged. We surmised that the neutral or positive effects of the technology on job opportunities during the first part of the computer era would not be characteristic of the next period.

We also started with the presumption that the nation's ability to make constructive use of the new technology would be determined in considerable measure by its ability to adjust its educational and labor market institutions to the accompanying employment disruptions that affect differentially the disadvantaged members of the labor force.

This chapter opens with a quick review of some of our major findings about the impact of the new technology on employment before drawing the lessons for policy and for future research. In much of our research we have found that employment changes are not the result of new technology alone but also of other potent economic forces, among the most important of which was the heightened competition in the United States and in international markets. Undoubtedly, renewed competition has encouraged and often forced U.S. employers to seek new ways of cutting costs, particularly labor costs.

A number of important impacts on employment can then be identified as follows:

- There is a pronounced tendency for the new technology to change the content of work at all levels. On one hand, the new technology permits a sharp reduction of paper-oriented work susceptible to being automated. This work may range from highly routine tasks, such as posting or transcribing data, filing, and carrying out repetitive calculations to relatively complex procedures requiring considerable judgment and skill. On the other hand, the new technology is making it possible to broaden the range of responsibility for low-level personnel and to provide opportunities for new initiatives in areas of management and control for middle- as well as upper-level personnel. In our view then, the new technology is not resulting in broad-scale deskilling of jobs. Rather, the reverse is true.

- There is a tendency for management to reexamine hiring standards in the face of this changing work content in order to add to its ranks only those with the educational and social qualifications deemed necessary to effectively perform the new tasks at hand. In many instances, these new hiring standards are interpreted as requiring at least a junior college degree.
- The substantial reductions in the costs of transmitting information, even over long distances, suggest that many employers in high-cost urban centers who cannot find adequate numbers of educated technical and clerical staffs will move many support operations to more attractive distant locations, even overseas.
- The long-term decline in the number of employees in the manufacturing sector, who now represent less than one in five of the total labor force, is likely to accelerate as advances in automation occur, including greater use of robots.
- Minorities who frequently fail to complete high school or, if graduated, fail to pursue college training are likely to be particularly affected by the upward drift in employers' hiring standards.
- Minority women, who have long been concentrated in our inner cities and who had made substantial gains in clerical occupations in recent years, are having to face the unhappy prospect of seeing many of these opportunities evaporate because of the slowing down of employment growth in back offices and the relocation of many of those jobs to outlying locations.
- Both within manufacturing and services, there is a weakening of internal channels of upward mobility. In the past many workers with limited educational credentials were able to pick up skills in the workplace and subsequently move up the job ladder through these internal channels.

Policy Implications: Training and Retraining

There is no agreement among the experts, employers, or the public at large about the minimum desirable level of educational preparation for the future members of the work force. Part of the difficulty lies in distinguishing between credentials and competences; part in agreeing on the criteria to measure competences. A middle-of-the-road position holds that future entrants into the work force should have at least the competences comparable to those achieved in twelve years of education, such as reading skills, arithmetic ability, and analytical capabilities. Currently, about one in six among sixteen- to nineteen-year-old youth fails to meet this standard. Among blacks, the shortfall is in the range of one in four, and among Hispanics about one in two. In recent years many states have become concerned about the shortcomings of their educational systems and have begun to institute remedial programs, which range from raising standards for promotion and graduation to revising curricula and improving teachers' salaries.

Unfortunately large improvements in student performance in the near or middle term are unlikely. As a result, it is becoming increasingly clear that those who are left behind by the educational system—those with less than a high school education—are likely to be left behind by the labor market. We are in danger of creating a class of permanently unemployable or "underemployable" individuals. There are, of course, jobs that still demand little skill, but it is doubtful that there will be enough of them to go around, given the strong tendency toward general upgrading of hiring requirements. There is nothing in the currently evolving transformation that leads us to expect a repetition of the late 1960s, when the economy ran at full speed and unemployment rates dropped to below 4 percent. In a manufacturing economy, almost anyone was employable if she or he was willing to expend physical effort. In a service economy this is not true.

The increasing disinclination of large employers to hire workers who have not earned a high school diploma and who do not have adequate competences in English, arithmetic, and analysis bodes ill for many inner-city youths, so many of whom drop out of high school before they acquire these competences. The gap between the meager credentials these young people are able to offer and what employers want is often too wide to be closed by short-term remedial programs.

The foregoing analysis should not be read as a lack of support for efforts to improve the performance of the basic educational system for all young people. Every community and every state—as well as the federal government—has a stake in speeding up the accomplishment of educational reforms for reasons of defense, economic well-being, and individual fulfillment. But until the long-term objective is achieved and everybody capable of being educated is schooled to a level that will enable them to become employable, it is essential that training and employment programs recognize that remedial educational opportunities must be part of their mission and goals.

Federal and state governments that underwrite employment and training programs should not focus on cost-benefit ratios if by so doing they reduce or eliminate opportunities for educationally deficient workers to improve their competences. The educationally deficient, especially among youth, have the highest potential returns from participating in successful remedial programs. The additional costs to bring them into the regular labor force will be small compared to their life-time earnings.

Many employers, especially large employers in the service sector, have been hiring young people who have at least a high school diploma and preferably a junior college degree partly because the new technology is developing so fast. Employers recognize that there will be frequent changes in equipment and systems to which their work force will have to adjust. Therefore, they seek employees who will be able and willing to be trained and retrained as the technology changes. In many cases, the required training will be relatively short and will be provided in-house. When large-scale adjustments must be made in how the work is structured and performed, selected groups of workers may be sent to external training courses for several weeks or even longer.

One of the changes we noted in the training environment is the extent to which many white-collar workers who are seeking career advancement are signing up for semester and year-long courses, usually at nearby community and senior colleges. Many employers pay for all or part of the tuition. One of the concomitants of the economy's steady shift from goods to services output is the greater tendency of workers who, in search of career mobility, participate in continuing classroom instruction in order to master the new techniques.

Increasing numbers of new workers are now entering the work force with some knowledge of and acquaintance with computers, having grown up in a world where computers are taken for granted and are no longer restricted to experts. But the opportunities to become "computer literate" while at school vary widely.

Inequalities in educational allocations are likely to limit severely the ability of many pupils in low-income neighborhoods to become computer-literate and to learn how to make use of computer-aided instruction, especially because additional funds are required to improve instruction in basic educational skills. These pupils will face additional handicaps when they enter the labor market and face competition with others who have had access to a richer curriculum that included training in the use of the computer. Our society faces the challenge of helping to prepare future workers more effectively to cope with a world in which computers will be pervasive.

A related question is the training and retraining of currently employed workers. Most employers underwrite the training their workers require to cope with new machines and new systems. In most instances, the costs of retraining are modest compared to the gains in efficiency and quality that the new systems make possible. Moreover, if employers seek to avoid these retraining costs, they are likely to find it difficult to hire qualified workers because in an environment of exploding technology the supply of competent workers often lags behind demand.

Still another dimension is the retraining of displaced workers in order to facilitate their reemployment. There is only scattered information available, which points in the following directions. If a displaced worker has had only a modest amount of schooling (less than high school graduation) and has no special aptitude for the skills that the computer requires, retraining is likely to prove difficult. This was a principal finding of the follow-up study among automotive workers in the Los Angeles area who had lost their jobs and had received substantial retraining assistance. Only about 5 percent made a successful transition into computer-connected employment. In addition, as noted earlier, young people entering the labor force will increasingly have the edge over displaced older workers, both because of a more positive orientation and a more extended exposure to the new technology.

Because workers' attitudes toward retraining will in large measure determine how well they succeed, the growing number of collective bargaining agreements being negotiated that provide significant funding for the retraining of future redundant workers, as between the UAW and GM or Ford, may suggest greater potentials than in the past. It would surely be premature to write off retraining as a significant adjustment mechanism.

There are two additional observations about the retraining issue that warrant brief discussion. From high-tech companies in the Silicon Valley and financial firms in large metropolitan centers to apparel manufacturing firms in New York City and auto manufacturers in the Detroit region, we found an increasing tendency among employers to externalize retraining through greater reliance on community colleges. This tendency presents interesting new opportunities to integrate better short-term retraining efforts into long-term training curricula. Some of the community colleges and specialized institutions that we visited—including the Fashion Institute of Technology, the College for Human Services in New York, Foothill Community College in the Silicon Valley, and the UAW-Ford Training Center in Detroit—were very much aware of the need to develop more coherent curricula to meet the long-term career needs of individual workers while at the same time to respond to demands from employers to offer training designed to overcome skill shortages.

The second observation relating to the retraining issue is that one should not conclude from the above that worker displacements, wherever they occur, reflect simply a firm's or an industry's lack of retraining efforts. No amount of retraining in the steel or auto industry can make up for the fact that shrinking markets, new technology, and increasing competition forced many of the firms in these industries to shrink dramatically in order to survive. Clearly, at some level macroeconomic policies (e.g., demand policy, tax policy, trade policy) play important roles in accelerating or retarding worker displacement, and it would be wrong to view technology as the sole culprit for the dramatic shrinkage of some of the older industries.

Up to this point, we have stressed problems confronting three groups of workers:

- Young people who fail to complete high school lack the foundations that would enable them to obtain employment in advanced service industries.
- Workers in the smokestack industries, mostly white and minority men (and some women), who have experienced, for a series of reasons only partly connected with the computer, continuing and, recently, accelerated declines in their jobs.
- Finally, adult clerical workers, both white and minority women, who in past decades were able to find jobs in the rapidly growing white- collar work force, are now threatened by the new technology in two respects: (1) the ability of the new machines to expand output without further increases in the number of employees and (2) the potential for employers to relocate many of these jobs away from metropolitan centers where the minority population is concentrated.

Between 1962 and 1981, the United States engaged in a potpourri of employment and training programs. In the late 1970s, for example, the government financed 750,000 CETA public service employment (PSE) jobs available to disadvantaged persons at or close to the poverty level. But the program lost favor with the public and the Democratic Congress, and was therefore scuttled by the Reagan Administration in its 1981 budget cutbacks.

If one starts with the premise that the work ethic is basic to the U.S. value structure, it follows that all our principal institutions— government, business, and labor—should address the problem of what they can do to assure that individuals who need and want to work have an opportunity to do so when there is a shortfall in private sector jobs. The fact that the CETA experience was uneven and that relatively few CETA workers were able to make the transition into regular employment does not support the conclusion that public job creation efforts should be permanently abandoned. Rather, such efforts are an essential ingredient of any comprehensive effort to assure that the labor force and labor markets perform better. A few illustrations will help to support this contention.

Young people who reach working age poorly prepared for employment need opportunities for remedial education, opportunities to acquire occupational skills, and opportunities for establishing an employment record. It is highly questionable that they could increase their employ-ability without developing a record of performance in a disciplined work environment.

At the opposite end of the age distribution are older displaced workers. Because they own their homes and have many local commitments and obligations, older displaced workers have limited or no geographic mobility. Those who have the chance to collect good retirement benefits based on "30 years and out" should make ends meet, especially if they obtain some part-time employment and/or if their spouses hold jobs. But others will be under monetary as well as psychological pressures if they lose their jobs in their early fifties, before they become eligible for Social Security. They, their families, and their communities would be better off if they could be productively employed in these interim years in public work when private work is not available.

Finally, there are many adults, both men and women, who will lose their jobs because of cyclical and structural changes in the economy and will not find another one in the six to nine months during which they are eligible for unemployment compensation. Once again, it would make more sense for them, their families, and their communities if they had the opportunity to be employed in publicly financed jobs until the private sector expands again.

It would take us too far afield to discuss the multiple arrangements that are required to have the public job program operate effectively, including wage levels and the need for close linkages with the Employment Service. But these and still other difficulties are minor when compared to the large-scale financial and societal losses that we currently endure because of a shortfall in jobs.

Directions for Future Research

Most recent research on the impact of technology on employment, including ours, has been based on micro case studies. Clearly, much has already been learned from such case studies, but there is a limit to how far one can pursue this approach regardless of how rich the materials or how suggestive the analyses. Economists must also be concerned with macro-trends, particularly with the efficiency with which a society makes use of all of its resources, human and physical. Exemplary behavior by a few organizations can provide models and cues, but it cannot tell us how fast or how well the economy as a whole is responding or will respond to new technological opportunities and challenges.

The studies on which we have reported in this book call attention to the following critical dimensions that appear to be interacting with the new technology: shifts in the location of business and of many of its critical functions within the United States and also abroad; shifts away from goods production in favor of services; the growing importance of international developments for the future of the U.S. economy; the large numbers of poorly educated persons, native-born and foreign-born, who face a labor market characterized by a steady increase in qualifications needed for regular employment in other than "dead end" jobs; the steady increase in the national level of unemployment from around 4 percent at the peak of expansion in the late 1960s to 7.5 percent in the mid–1980s; and the much higher average rates of unemployment that are characteristic of youth and minorities.

We believe that, with the foregoing as signposts, a useful way to move from micro case studies to a broader macro understanding is by analyses of adjustment and transformation in metropolitan labor markets. In particular, metropolitan labor markets are a good place to explore which groups of workers are gaining and which are losing as the new technology proceeds to change jobs, skill requirements, entry-level requirements, and other labor market characteristics. Furthermore, metropolitan labor markets provide the proper scale at which to assess the influence of the new technology on the geographical restructuring of employment, which is reflected in suburbanization, regionalization, and even the internationalization of jobs.

The metropolitan economies also provide a useful environment for investigating an issue that has heretofore received inadequate attention: the changing relative importance of small and large firms in the labor market and the implications that the changing relations between them have for employment opportunities. In recent years, there has been considerable evidence that small firms have been responsible for the major share of job growth, although there is little indication that large corporations are playing a reduced role in terms of assets controlled or goods and services produced. Because small organizations have traditionally tended to be wage-followers rather than wage-setters and to have weaker internal labor markets than their larger counterparts, it is important to learn more regarding the significance of their apparently increasing weight in metropolitan labor markets.

Another concern that can be addressed by metropolitan studies relates to immigrants. Most large cities are major points of entry for large flows of immigrants whose absorption into the U.S. economy poses special problems. For example, "Anglos" in Los Angeles have become a minority, and more than one out of every four New York City residents is from abroad. One important objective of any metropolitan study of immigrants would be to identify the variety of ways that different groups from Central and South America, the Far East, or Southeast Asia make use of a wide range of resources, from financial assets to personal contacts, community networking, and access to public amenities, to gain a toehold and make progress in the economy. The range of "self-help" mechanisms to which some of these groups resort should be studied in the hope of identifying those that might be useful to native groups who are experiencing difficulties in their job and career efforts.

The United States hosts large flows of new immigrants who are at best ambivalent about and at worst hostile to learning and using English as their primary tongue. The language-job-career relationship warrants attention, particularly as the economy continues to tilt in the direction of activities for which English language skills are a prerequisite to employment and career progression. In addition, many large U.S. cities are characterized by large numbers of high school dropouts. Some cities have done more than others to pursue remedial efforts in both the government and voluntary sectors—efforts that should be assessed prior to replication.

Alternatives to regular employment are being pursued by disadvantaged youth and adults in the cities, ranging from street-vending and other off-the-record employment (often without benefits or employment taxes) to illicit and illegal activities. It would be useful to learn more from the known data about these "gray areas" and about their contribution to the employment and career opportunities for disadvantaged persons.

Employment opportunities that would enable the disadvantaged to obtain entry-level jobs in either public agencies or large corporations have declined as a result of a number of structural changes, some of which have been identified in this and earlier chapters. Furthermore, we have learned from some of the research presented here that these new trends have had differing effects on different groups, but particularly negative impacts on black and Hispanic males. Beyond this, little information has been garnered about what the changing nature of job openings and career advancement implies for these groups and what can be done to assist them to meet current and prospective changes.

We called attention earlier to the dramatic changes that are resulting from the reorganization of work and the implications of such changes for the long-term education-work linkage. On this front, and in addition to our concern about primary and secondary schooling, there is a need to investigate in greater detail the new and, we believe, closer integration between community colleges and rapidly changing metropolitan labor markets. Here again, research focusing on key metropolitan areas would appear to offer a promising strategy in learning how to cope better with the problems associated with the impact of technology on work and career opportunities.

About the Authors

Eli Ginzberg is director, Conservation of Human Resources, and A. Barton Hepburn Professor Emeritus of Economics, Columbia University. He is the author of numerous books on manpower economics and related subjects, including *Understanding Human Resources: Perspectives, People and Policy* (1985), and coauthor of *From Health Dollars to Health Services: New York City 1965–1985* (1986) and *Beyond Human Scale: The Large Corporation at Risk* (1985).

Thierry J. Noyelle is senior research scholar, Conservation of Human Resources, Columbia University. He is the author of several books, including *Beyond Industrial Dualism: Market and Job Segmentation in the New Economy* (Westview, forthcoming) and coauthor of *Services/The New Economy* (1981) and *The Economic Transformation of American Cities* (1984).

Thomas M. Stanback, Jr., is adjunct senior research scholar, Conservation of Human Resources, Columbia University, and professor of economics, New York University. He is the author of several books, including *Understanding the Service Economy* (1979), and coauthor of *Services/The New Economy* (1981), and *The Economic Transformation of American Cities* (1984).

Index